Physics, Chemistry and Technology of Solid State Gas Sensor Devices

CHEMICAL ANALYSIS

A SERIES OF MONOGRAPHS ON
ANALYTICAL CHEMISTRY AND ITS APPLICATIONS

Editor

J.D. WINEFORDNER

VOLUME 125

A WILEY-INTERSCIENCE PUBLICATION

JOHN WILEY & SONS, INC.

New York / Chichester / Brisbane / Toronto / Singapore

Physics, Chemistry and Technology of Solid State Gas Sensor Devices

ANDREAS MANDELIS

Photothermal and Optoelectronic Diagnostics Laboratory
Center for Hydrogen and Electrochemical Studies
Department of Mechanical Engineering
University of Toronto
Toronto, Ontario, Canada

and

CONSTANTINOS CHRISTOFIDES

Department of Natural Sciences
University of Cyprus
Nicosia, Cyprus

A WILEY-INTERSCIENCE PUBLICATION

JOHN WILEY & SONS, INC.

New York / Chichester / Brisbane / Toronto / Singapore

Chem
TP
754
.M36
1993

This text is printed on acid-free paper.

Copyright © 1993 by John Wiley & Sons, Inc.

All right reserved. Published simultaneously in Canada.

Reproduction or translation of any part of this work beyond
that permitted by Section 107 or 108 of the 1976 United
States Copyright Act without the permission of the copyright
owner is unlawful. Requests for permission or further
information should be addressed to the Permissions Department,
John Wiley & Sons, Inc., 605 Third Avenue, New York, NY 10158-0012.

Library of Congress Cataloging in Publication Data:
Mandelis, Andreas.
    Physics, chemistry and technology of solid state gas sensor devices / Andreas
    Mandelis and Constantinos Christofides.
        p.   cm. — (Chemical analysis : v. 125)
    "A Wiley-Interscience publication."
    Includes bibliographical references and index.
    ISBN 0-471-55885-0 (alk. paper)
    I. Gas-detectors. I. Christofides, Constantinos. II. Title.
III. Series.
TP754.M36   1993
681'.2—dc20   93-6565

Printed in the United States of America

10 9 8 7 6 5 4 3 2 1

To Nancy and Loukia

CONTENTS

PREFACE

Since the days of the industrial revolution humans have been living in a continuously changing world. During the early industrial era all technological developments relied on sets of conflicting criteria, the nature of which was dictated by the socioeconomic forces of the time and was often consistent with the desire for scientific progress.

In today's world two additional factors appear in defining the state of technological development: health and environment. These new factors are not simply in the form of societal demands but in many ways represent historical necessities, apparently stemming from increasing feelings of self-preservation among the human race in the face of adverse environmental effects created by short-sighted technological activities. Never in the recorded historical past did the planet face the magnitude of health and environmental issues confronting us today.

The extensive pollution problems in our societies are adversely affecting our health, and therefore any new technological impetus must incorporate in its theme any and all safeguards contributing to environmentally sound activities. As a result, new and powerful research areas have emerged in our battle for awareness and environmental and health monitoring. One such area is that of solid state gas sensor research and development. It started out as uncoordinated individual efforts of a few investigators in the 1950s and 1960s and has blossomed into a multidisciplinary international research field since then. A growing number of laboratories around the globe are introducing novel methodologies and devices to address specific needs associated with relevant technological developments. Several thousand articles have appeared in archival journals, as well as in more popularized publications, depending on the intended audience, all in a time span of little more than two decades (1970 to the present).

A few books have also appeared with valuable information on a variety of subjects regarding solid state sensors. Among those, we wish to acknowledge the impact of the following publications: Jiri Janata and Robert J. Huber, eds., *Solid-State Chemical Sensors*, Academic Press, Orlando, Florida, 1985; P. T. Moseley and B. C. Tofield, eds., *Solid State Gas Sensors*, Adam Hilger,

xiii

Bristol, England, 1987; Marc J. Madou and S. Roy Morrison, *Chemical Sensing with Solid State Devices*, Academic Press, San Diego, 1989. Last but not least, an invaluable series of proceedings of international conferences on solid state sensors and actuators, under the name *Transducers (19...)* has been contributing to the tremendous pace of solid state sensor development in the international research community.

It has been apparent to us for some time that the fast pace of developments in this field has barely allowed books and monographs to focus on the sensitivity and performance of various sensors, with very little review and coordination work yet published on the sensor science and technology as a whole. As a result, we decided to focus in this volume on the physics, chemistry and technology of the devices, i.e., on the principles of solid state sensor operation, which in turn expose the underlying rich fabric of the inter-disciplinary science that governs modern sensing devices.

In that sense, we hope that the present volume will act as a unifying focus in the presentation of the scientific principles of sensor operation.

The widely diverse material on which we have drawn has been grouped into coherent pedagogical families of principles, and thus it is hoped that it will be suitable for classroom use, as well as a guide to professional researchers and users of solid state sensor devices. We have specifically focused on fabrication technology, performance of devices, areas of application, and integration/multiplexing trends in sensors. In the chapters of this volume, these topics usually follow the description of the fundamental physical or chemical principles of sensor operation.

In closing, we would like to thank various fellow researchers in the area of solid state sensors who kindly provided us with figures of their devices, subsequently used in this volume. Among them we count Dr. M.A. Butler and Dr. J. N. Zemel. Special thanks are due to all other colleagues who have kindly updated us on recent developments of their research. We truly apologize to those authors whose work we may have inadvertently omitted.

We are also thankful to the secretarial personnel of the Department of Mechanical Engineering at the University of Toronto, Ms. W. Smith, Mrs. G. Néné, Ms. R. Leo, and Mrs. M. Tompsett, for carrying out the task of putting the text into a correct, coherent format. Among others, we acknowledge the continuing support of the Center for Hydrogen and Electrochemical Studies and the Ministry of Energy, Mines and Resources Canada, in the encouragement they have given us over the years to pursue our research efforts in the exciting and rewarding field of photothermal and solid state hydrogen sensors.

Finally, our deepest thanks to our wives Nancy and Loukia for all the moral support and wonderful patience during the writing of this work.

ANDREAS MANDELIS
CONSTANTINOS CHRISTOFIDES

Toronto, Canada
Nicosia, Cyprus
September 1993

LIST OF ACRONYMS

ADFET	Adsorption field effect transistor
ALW	Antisymmetric Lamb waves
CCD	Charge-coupled device
CFT	Charge-flow transistor
$C-V$	Capacitance–voltage
CW	Continuous wave (laser)
DIP	Dual in-link package
DLTS	Deep-level transient spectroscopy
EPS	Electron photoconducting spectroscopy
FET	Field effect transistor
FOS	Fiber-optic sensor
FTIR	Fourier transform infrared
GC	Gas chromatograph(ic)
HID	Hydrogen-induced drift
HV	High vacuum
IDT	Interdigital transducer
IGFET	Insulated-gate field effect transistor
IR	Infrared
ISFET	Ion-sensitive field effect transistor
$I-V$	Current–voltage
$J-V$	Current density–voltage
M-S	Metal semiconductor
MiS	Metal–(thin) insulator–semiconductor
MIS	Metal–insulator–semiconductor
MISFET	Metal–insulator–semiconductor field effect transistor
MNOS	Metal–nitride–oxide–semiconductor
MOS	Metal–oxide–semiconductor
MOSFET	Metal–oxide–semiconductor field effect transistor
MOST	Metal–oxide–semiconductor transistor
NA	Numerical aperture
OGFET	Open-gate field effect transistor
PA	Photoacoustic
$P-C$	Pressure–concentration
PD	Photodiode
PE	Pyroelectric
PGA	Pyroelectric gas analyzer

PLS	Partial-least-square
PMT	Photomultiplier tube
PMW	Plate-mode wave
ppb	Parts per billion
PPE	Photopyroelectric
ppm	Parts per million
ppmv	Parts per million by volume
ppt	Parts per trillion
PQCMB	Piezoelectric quartz crystal microbalance
PRESSFET	Pressure field effect transistor
PT	Photothermal
PVDF	Poly(vinylidene fluoride)
rf	Radio-frequency
RH	Relative humidity
rms	Root mean square
SAFET	Surface-accessible field effect transistor
SAW	Surface acoustic wave
SFR	Spin-flip Raman
SGFET	Suspended-gate field effect transistor
SIMS	Secondary ion mass spectrometry
SLW	Symmetric Lamb waves
SPR	Surface plasmon resonance
STP	Standard temperature and pressure
STW	Surface transverse waves
TMOS	Thin metal–oxide–semiconductor
TMOSFET	Thin metal–oxide–semiconductor field effect transistor
UHV	Ultrahigh vacuum
UV	Ultraviolet
VHT	Very high temperature
XIS	Conductor (electronic or ionic or a gas)–electrical insulator–semiconductor
XPS	X-ray photoelectron spectroscopy
YSZ	Yttria-stabilized zirconia

CHEMICAL ANALYSIS

A SERIES OF MONOGRAPHS ON
ANALYTICAL CHEMISTRY AND ITS APPLICATIONS

J. D. Winefordner, *Series Editor*
I. M. Kolthoff, *Editor Emeritus*

CHAPTER

1

INTRODUCTION

1.1. WHY GAS SENSORS?

Today's modern industrialized society has brought to the world numerous goods and services, as well as a series of problems related to technological development. Ever-increasing industrialization makes it absolutely necessary to constantly monitor and control air pollution in the environment, in factories, laboratories, hospitals, and generally technical installations. The field of chemical sensors continues to be a topic of interest in the United States and Canada (Zemel, 1987), in Japan (Yamazoe, 1987), and in Europe (Jones, 1987).

In recent years, several types of gases have been used in different areas. In fact, in many industries gases have become increasingly important as raw materials (Vaz de Campos, 1987). For this reason among others, it has become very important to develop highly sensitive gas detectors to prevent accidents due to gas leakages, thus saving lives and equipment. Such detectors should allow continuous monitoring of the concentration of particular gases in the environment in a quantitative and selective way. The following list gives both constraints and requirements for an "ideal" chemical detector (Christofides and Mandelis, 1990; Graber et al., 1990):

a. Chemically selective
b. Reversible
c. Fast
d. Highly sensitive
e. Durable
f. Noncontaminating
g. Nonpoisoning
h. Simple operation
i. Small size (portability)
j. Simple fabrication
k. Relative temperature insensitivity

1

l. Low noise

m. Low manufacturing costs

In addition, this control system should be financially accessible to potential users. Another area in which gas detectors are also very useful is the field of surface science. In fact, devices that operate on the principle of a decrease of the work function of the selective metal by adsorption of gas are the main means for the study of the surface–gas interaction in the metal–gas system.

1.2. ORGANIZATION OF THE BOOK

This book presents the development, history, theoretical basis, and experimental performance of solid state gas detectors under flow-through conditions reported to date, such as semiconductor, photothermal, optical and fiber-optic-based, piezoelectric (bulk and acoustic wave), pyroelectric, and thermal devices. The main emphasis of the presentation is in the physics and chemistry of these devices. Non-solid-state gas sensors such as the flame ionization detector are beyond the scope of this volume. For a review of these devices see Kings (1970).

Some key points concerning gas–surface interactions are discussed in Chapter 2. It is well known that the role of catalysis is very important in the field of gas sensor technology. In that chapter very useful information can be found concerning the adsorption, absorption, and desorption of gas molecules in catalytic surfaces. A special section concerning the adsorption properties of noble metals has also been included.

The purpose of Chapter 3 is to review progress made in the vast field of semiconductor-based gas sensors. The reader is offered the fundamental physics and a critical review and classification of several types of semiconductor gas-sensing devices. Attention has been given to semiconductor devices that have been used under flow-through conditions (in air or in an inert atmosphere) as representative of possible environmental and pollution monitoring applications.

The concepts, principles, and performance of novel photonic and photoacoustic gas sensor devices are discussed in Chapter 4.

Chapter 5 deals with fiber-optic devices. As in the previous chapter, the extensive theoretical section that discusses the basic operations of the devices is followed by several experimental results obtained to date.

In Chapter 6 we describe the fundamental theory of the quartz crystal, as well as the theory of operation of the piezoelectric crystal microbalance. Several tables present important information concerning the performance of the piezoelectric quartz crystal microbalance.

Chapter 7 is dedicated to surface acoustic devices such as Rayleigh surface acoustic wave sensors, plate-mode structures, and transverse wave elements.

Chapter 8 presents pyroelectric and novel thermal gas sensors. Special attention has been paid to the theory of these devices, as they are relatively new with fundamental questions remaining open.

Finally, Chapter 9 consists of general conclusions and also presents a few ideas concerning future prospects in the field of solid state gas sensors.

A critical comparison of various hydrogen-sensing devices is presented in the Appendix (see page 313). This is the widest, broadest, and most researched device family, members of which span almost all device categories in this book. Therefore, direct comparisons of the performance of such devices can be readily made only in the H_2-sensing case. The room temperature performance of these devices has been highlighted. A useful feature of the Appendix is a comparison of operating characteristics of each device (see Table A.1).

REFERENCES

Christofides, C., and Mandelis, A. (1990). *J. Appl. Phys.* **68**, R1.

Graber, N., Lüdi, H., and Widmer, H.M. (1990). *Sens. Actuators* **B1**, 239.

Jones, T.A. (1987). *Proc. Symp. Chem. Sensors, Electrochem. Soc.* **87-9**, 12.

Kings, W.H. (1970). *Environ. Sci. Technol.* **4**, 1136.

Vaz de Campos, E.F.P. (1987). *Int. J. Hydrogen Energy* **12**, 847.

Yamazoe, N. (1987). *Proc. Symp. Chem. Sensors, Electrochem. Soc.* **87-9**, 1.

Zemel, J.N. (1987). *Proc. Symp. Chem. Sensors, Electrochem. Soc.* **87-9**, 23.

INTERACTIONS OF GASES WITH SURFACES: THE H_2 CASE

2.1. INTRODUCTION TO CATALYSIS

This section describes the role of catalysis in the field of gas sensor technology. Two very important cases can be distinguished: (a) calorimetric detection, which measures gas concentration vs. temperature rise produced by the heat of reaction on a catalytic surface; and (b) detection due to the change of electrical parameters, such as the change in electrical conductivity induced by adsorption or reaction of gases on the solid surfaces. The above two cases have been extensively reviewed by Gentry and Jones (1986).

Catalytic effects play an important role in the field of gas detection. Solid-state gas sensors are directly related to the phenomenon of catalysis. Catalytic processes not only control the rate at which a chemical reaction approaches equilibrium (this considerably affects the response time in the case of gas detection) but also affect sensitivity and selectivity. The ideal catalyst is one which increases the rate of the gas–surface interaction without itself becoming permanently affected by the reaction. Thus, the response time will be fast and the process will be reversible, and finally the sensor will possess three important properties (see Chapter 1): speed, durability, and reversibility.

Figure 2.1 (Gentry and Jones, 1986) shows schematically the energetics of the reaction (both catalyzed and uncatalyzed):

$$A + B \rightleftarrows C + D \tag{2.1}$$

It can be seen that the uncatalyzed reaction is characterized by an extremely high activation energy, E_g. In the catalyzed reaction the gaseous species A and B adsorb on the surface with an exothermic heat of adsorption ΔH (state I). The state I is followed by state II toward the reaction products C and D (state II) characterized by the activation energy E_c, which is much lower than E_g. It is important to note that the activation energy is not the only factor determining the activity of the catalyst. The following five factors play a role in affecting catalytic reactions (Gentry and Jones, 1986):

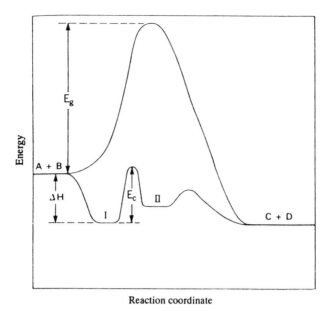

Figure 2.1. Energy configuration for the reaction of Eq. (2.1). Activation energies: E_g, homogeneous gas phase reaction; E_c, heterogeneous catalyzed reaction (Gentry and Jones, 1986).

a. Transport of gases to the solid surface

b. Adsorption of the gases on the solid surface

c. Reaction between the adsorbed species and/or with the solid surface

d. Desorption of the surface species and products of surface reactions

e. Transport of the gaseous reactants and products away from the surface

2.2. PHYSISORPTION AND CHEMISORPTION

As is well known, gas adsorption on surfaces occurs because the atoms or ions at the surface of the solid cannot fully satisfy their valency or coordination requirements. This leads to a certain permanent force acting inward. Then, the adsorption of external species that happen to be in the neighborhood of the surface reduces the surface energy of the solid (Somorjai, 1972). One can distinguish two types of adsorption. (a) *Physisorption*: weak attraction followed by gas adsorption due to van der Waals forces; physisorption is characterized by a low heat of adsorption. (b) *Chemisorption*: in the case of high surface energy the gas may become adsorbed through an exchange of electrons with the surface, that is, chemical bonds are formed; the heat of chemisorption is higher than that of physisorption. Table 2.1 presents some typical heats of adsorption for various gases and metals (Somorjai, 1972). As is shown in Table 2.1, the heats of chemisorption are in the range of 15 to 200 kcal/mol, whereas physical adsorption is only up to 15 kcal/mol.

In sensor technology two types of catalytic materials are mostly used: metals and semiconductor metal oxides (Gentry and Jones, 1986). In metals, the electrons involved in chemisorption are the free electrons of the imcomplete d-band. Thus, one can understand why the noble metals (Pt, Pd, Rh, Ir) are the most active catalysts (Kohl, 1990), since they possess a partially filled d-band. On the other hand, the semiconductor metal oxides contain lattice defects due to an excess or deficit of oxygen in the lattice. The association of electrons with these defects following chemisorption allows a certain change of the electrical conductivity of the oxide (Keramati and Zemel, 1978).

Table 2.1. Heats of Chemisorption of O_2, H_2, N_2, and CO on Several Metal Surfaces

Gas	Material	ΔH_{ads} (kcal/mol)
O_2	W	194
	Mo	172
	Rh	118
	Pd	67
	Pt	70
H_2	Ta	45
	W	45
	Cr	45
	Mo	40
	Ni	30
	Fe	32
	Rh	28
	Pd	26
	Mn	17
N_2	W	95
	Ta	140
	Fe	70
CO	Ti	153
	W	82
	Ni	42
	Fe	46

Source: Somorjai (1972).

In the case of metals there is also a certain change of electrical conductivity, but the relative change is very small owing to the high concentration of conduction electrons and is therefore difficult to measure.

In order to have successful gas sensors two main properties concerning the catalyst are necessary: (a) in the case where the detectivity of the sensor is due to the electrical conductivity shift, these charges must be proportional to the concentration of the detecting gas; and (b) in the case of calorimetric detection the heat of adsorption must be independent of the coverage. If these two properties are satisfied, then the sensor may possess a linear response, which is one of the most important properties of a gas detection device. The linear response occurs especially in the range of low concentration and not near the saturation level.

2.3. THE CHANGE OF WORK FUNCTION

The adsorption of atoms or molecules on a metal surface changes the distribution of charges and gives rise to changes in the work function, ϕ (Riviere, 1970). The formation of adsorbed negative ions or the ionization of atoms (transfer of electrons into the surface) decreases or increases the work function, respectively. Thus, the monitoring of ϕ of the known metal surface under investigation can give important information concerning the nature of the gas adsorption on the surface as well as information concerning the gas–surface interactions and the kinetics of the adsorption–absorption process. In fact, very often the work function change is directly related to surface coverage. Several experimental techniques have been used since 1940 for measuring work function changes upon chemisorption, such as retarding potential (Gysae and Wagener, 1940), vibrating capacitor (Mignolet, 1950), and field emission (Gomer, 1953). Table 2.2 presents the work function changes of metals, $\Delta\phi$, upon the adsorption of various gases. In Chapter 3 a complete review can be found of semiconductor gas sensors, which operate by monitoring the work function shifts of catalytic metals.

2.4. THE PALLADIUM–HYDROGEN SYSTEM

It is well known that Pd has high hydrogen solubility (Lewis, 1967). The absorption of hydrogen by Pd during electrolysis was observed in 1868 by Thomas Graham (see Lewis, 1967). Since then, the palladium–hydrogen system has been studied extensively. Because of its selectivity to hydrogen absorption, Pd has been employed as a filter for hydrogen purification and

Table 2.2. Work Function Changes of Metals upon Chemisorption of Several Gases

Metal	Absorbing Gas	Work Function Change (eV)[a]
W	H_2	0.48
Fe	H_2	0.45
Ni	H_2	0.35
Cu	H_2	0.35
Ag	H_2	0.35
Au	H_2	0.18
Pt	H_2	0.14
W	O_2	1.19
Ni	O_2	1.60
Pt	O_2	1.20
Cu	O_2	0.68
Fe	CO	−1.50
Co	CO	−1.48
Ni	CO	−1.35
W	CO	−0.86
Pt	CO	0.18
Cu	CO	0.30
Ag	CO	0.31
Au	CO	0.92
Pt	C_2H_2	−1.40
Pt	C_2H_4	−1.11
Pt	C_3H_6	−1.36

Source: Somorjai (1972).

[a]Data from Gysae and Wagener (1940) and Mignolet (1950).

has also been used to provide hydrogen selectivity for various hydrogen detectors.

Banerjee and Lee (1979) developed a diffusion model for the hydrogen–palladium system, and Behm et al. (1980) studied the adsorption of hydrogen on Pd⟨100⟩ surfaces. The mechanism of hydrogen desorption has also been studied by Bucur et al. (1976). An extensive theoretical study on the chemisorption of hydrogen on palladium has been published by Nakatsuji et al. (1987). Equilibrium and kinetic measurements on thin Pd–H layers were performed by Bucur and Mecea (1980), Davenport et al. (1982), and Hong and Sapru (1987). Engel and Kuipers (1979) have used molecular beams to investigate the hydrogen scattering from, adsorption on, and absorption in Pd⟨111⟩. Qian and Northwood (1988) published a critical review of the experimental observations and theoretical models on hysteresis in metal–hydrogen systems. A thermodynamic study of the Pd–H$_2$ system at high temperatures and pressures was made by Picard et al. (1978). Special attention has also been paid to adsorption isotherms of hydrogen in the α- and β-phases of the H$_2$–Pd system (Simons and Flanagan, 1965).

The absorption of hydrogen in Pd depends on temperature and hydrogen concentration. Figure 2.2 shows a schematic comparison of solubility of hydrogen in Ni, Pd, and Pt metals at a pressure of 1 atm as a function of temperature (Lewis, 1967). A comparison between Pd and Pt as gates in metal oxide semiconductor (MOS) devices has also been made by Armgarth et al. (1982). These authors have shown that Pd is superior as a gate material for the detection of small amounts of hydrogen at room temperature. They also showed that Pt is more suitable for high hydrogen concentrations. From Fig. 2.2 one can easily see that Pd is predominant in hydrogen sensor technology owing to the much higher solubility of hydrogen in Pd than in Ni or Pt. In order to discuss a hydrogen detector response as a function of concentration (or hydrogen partial pressure), it is necessary to recall the surface–gas interaction in the hydrogen–palladium system. Gas–surface interactions are important as energy transfer mechanisms at the gas–solid interface on an atomic scale; so is the adsorption–absorption–desorption process.

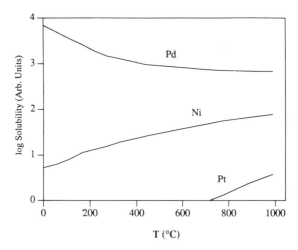

Figure 2.2. Schematic comparison of solubility of hydrogen in Pd, Ni, and Pt at a pressure of 1 atm as a function of temperature (Lewis, 1967).

The various experimental results dicussed in this book concerning H_2 detection have usually been interpreted as a function of the hydrogen concentration (or hydrogen partial pressure). In the hydrogen–palladium system the flux of the surface impinging molecules, Φ_0, of a particular gas is proportional to the partial pressure of the gas, p_g, the gas molecular weight, M_0, and the absolute temperature, T, assuming low enough concentrations and no interference from other ambient gases. According to the kinetic theory of gases the flux of gas molecules is given in Eq. (2.2) (Somorjai, 1972; Thomas and Thomas, 1967):

$$\Phi_0\left(\frac{\text{molecules}}{\text{cm}^2\cdot\text{s}}\right) = \frac{N_A p_g}{(2\pi M_0 RT)^{1/2}} \approx 3 \times 10^{22} \frac{p_g(\text{Pa})}{(M_0 T)^{1/2}} \qquad (2.2)$$

where N_A is the Avogadro number and R is the universal gas constant.

The majority of investigators in the field of hydrogen sensors interpret their results as a function of a controlled known experimental parameter, such as the molecular hydrogen partial pressure $p(H_2)$. For this purpose the atomic hydrogen pressure in the trapping medium must be given as a function of molecular pressure in the gas phase. The adsorbed hydrogen gas molecules are dissociated on the catalytic metal surface (usually Pd), and the H atoms are absorbed into the Pd bulk. Some of these absorbed atoms diffuse to the metal–substrate interface. According to Lundström (1981), equilibrium exists between the number of adsorbed H on the surface and that at the back interface (see Chapter 3).

2.4.1. The Langmuirian Model in an Inert Atmosphere

In the case of an enert atmosphere (such as N_2 or Ar), the only reaction taking place on the metal surface is (Lundström, 1981; Macly, 1985)

$$H_2 \underset{c_1}{\overset{d_1}{\rightleftharpoons}} 2H_a \qquad (2.3)$$

This reaction represents the dissociation of hydrogen on the Pd catalyst; H_a is the adsorbed hydrogen atom, and c_1 and d_1 are the rate constants of the reaction. After the absorption of hydrogen into and diffusion through the Pd film, another reaction takes place (Lundström, 1981):

$$H_a \underset{c_e}{\overset{d_e}{\rightleftharpoons}} H_b \underset{c_i}{\overset{d_i}{\rightleftharpoons}} H_{ai} \qquad (2.4)$$

where c_e, c_i, d_e, and d_i are rate constants; H_b and H_{ai} are the atomic hydrogen in the bulk and at the back interface, respectively. Assuming that the number of hydrogen adsorption sites per unit area on the Pd surface is N_e and that on the Pd–substrate interface is N_i, one can write (Lundström, 1981):

$$\frac{n_i}{N_i - n_i} = \frac{d_i}{c_i}\left(\frac{n_b}{N_b - n_b}\right) = \frac{c_e d_i}{d_e c_i}\left(\frac{n_e}{N_e - n_e}\right) \tag{2.5}$$

where N_b is the number of absorption sites in the bulk (per unit volume × thickness) of the Pd film; n_i n_b, and n_e are the concentrations of the adsorbed hydrogen atoms on the Pd–substrate interface, bulk, and surface, respectively. At equilibrium the forward and backward rates in Eq. (2.3) are equal, so that

$$c_1[H_2] = d_1[H_a]^2 \tag{2.6}$$

Furthermore, from Eqs. (2.3), (2.5), and (2.6) (Lundström, 1981),

$$\frac{n_e}{N_e - n_e} = \left[\frac{c_1}{d_1}p(H_2)\right]^{1/2} \tag{2.7}$$

Note that the partial hydrogen pressure, $p(H_2)$, has been introduced in Eq. (2.7), which leads to a more direct interpretation of a number of experimental results, expressed as a function of hydrogen partial pressure. Thus, it is necessary to introduce the *coverage* of hydrogen at the surface, $\Theta_e = n_e/N_e$, and at the Pd–substrate interface, $\Theta_i = n_i/N_i$. By combining Eqs. (2.5) and (2.7) and by taking into account the definition of the coverage one can write:

$$\frac{\Theta_i}{1 - \Theta_i} = A_H \frac{\Theta_e}{1 - \Theta_e} = K[p(H_2)]^{1/2} \tag{2.8}$$

where $A_H \equiv c_e d_i/d_e c_i$ and K is a constant $[K \equiv (c_1/d_1)^{1/2}]$ (Lundström, 1981) that depends mainly on the difference in adsorption energies at the surface and interface, respectively; K is given by the relation (Somorjai, 1972):

$$K(T) = 2.5 \times 10^{-5}\left[\frac{A_r}{TM_0}\right]^{1/2}\exp\left(-\frac{\Delta E_a}{2RT}\right) \quad [Pa^{-1/2}] \tag{2.9}$$

where A_r is a temperature constant (given in units of gK) and ΔE_a is the heat of adsorption "per molecule," i.e., per two atoms, if the adsorption is dissociative such as in the case of the Pd–H$_2$ system. The value of ΔE_a is approximately 1 eV (Lundström, 1981; Conrad et al., 1974). Finally the coverage of hydrogen, Θ_i, can be written as a function of the hydrogen partial pressure, $p(H_2)$ (Somorjai, 1972; Thomas and Thomas, 1967; Daniels and Alberty, 1961):

$$\Theta_i = \frac{K(T)[p(H_2)]^{1/2}}{1 + K(T)[p(H_2)]^{1/2}} \tag{2.10}$$

According to Lundström (1981), a linear relationship may be assumed between the measured response signal, $\delta\xi$ (which can be voltage, pyroelectric signal, frequency shift, optical signal, etc.), and the coverage of hydrogen atoms, Θ_i, under the condition $p(H_2) < 200$ Pa:

$$\delta\xi = \delta\xi_{max}\Theta_i \tag{2.11}$$

where $\delta\xi_{max}$ is the maximum signal response of the hydrogen detector corresponding to complete surface coverage (saturation), and $\Theta_i = \Theta_{is}$ at the saturation level (i.e., $t \to \infty$). Then, Eq. (2.11) may be written:

$$\delta\xi_s = \delta\xi_{max}\Theta_{is} \tag{2.12}$$

Using Eqs. (2.10) and (2.11), it is easy to rewrite Eq. (2.12) in the form of the Langmuir isotherm in the saturation regime:

$$\delta\xi_s = \delta\xi_{max}\left\{\frac{K(T)[p(H_2)]^{1/2}}{1 + K(T)[p(H_2)]^{1/2}}\right\} \tag{2.13}$$

Equation (2.13) is valid only in the case where the palladium layer is clean (not oxidized) and exposed to pure hydrogen (Thomas and Thomas, 1967).

2.4.2. The Langmuirian Model in the Presence of Air

A good example of the Langmuirian model for surface gas adsorption is illustrated in the case of the quartz crystal microbalance sensor. It has been hypothesized (Abe and Hosoya, 1984) that the ambient oxygen plays an important role in the Pd piezoelectric quartz crystal microbalance response: Preadsorbed O_2 on the Pd surface reacts with the introduced H_2 gas and forms H_2O, which leaves the Pd surface via evaporation. Vannice et al. (1970) have studied the absorption of H_2 and O_2 on platinum black. The great similarity between Pt and Pd allows a reasonable comparison to be made between the results of Vannice et al. (1970) and the hypothesis put forth by Abe and Hosoya (1984). According to the aforementioned authors, the water formed during the reaction leaves the surface and, in doing so, is not replaced by additional hydrogen. The following empirical chemical reactions describe the possible mechanism on the Pd surface (Lundström et al., 1989): dissociation–recombination of oxygen,

$$O_2 \leftrightarrows 2O_a \qquad (2.14)$$

dissociation-recombination of hydrogen,

$$H_2 \leftrightarrows 2H_a \qquad (2.15)$$

and water formation (Case 1),

$$O_2 + 2H_a \rightarrow 2OH_a \qquad (2.16a)$$

$$OH_a + H_a \rightarrow H_2O_{(g)} \qquad (2.16b)$$

or, equivalently, water formation (Case 2),

$$O_a + H_a \rightarrow OH_a \qquad (2.17a)$$

$$OH_a + H_a \rightarrow H_2O_{(g)} \qquad (2.17b)$$

In Case 1 according to Lundström et al. (1989) one can write:

$$\frac{\Theta_i}{1 - \Theta_i} \propto \left[\frac{p(H_2)}{p(O_2)} \right]^{1/2} \qquad (2.18)$$

If water production as in Case 2 follows and this water production is less than the oxygen recombination, then one can write:

$$\frac{\Theta_i}{1 - \Theta_i} \propto \left[p(H_2) \left(\frac{1 - \Theta_O}{\Theta_O} \right) \right]^{1/2} \propto \left[\frac{p(H_2)}{\sqrt{p(O_2)}} \right]^{1/2} \qquad (2.19)$$

where Θ_O is the coverage of oxygen atoms. According to Ponec et al. (1969), the reaction between hydrogen and oxygen at the Pd surface at $0\,°C$ goes to completion producing desorbable water, in agreement with the last stage, Reaction (2.16) or (2.17), of the four-step mechanism, Eqs. (2.14)–(2.18). However, according to Lundström et al. (1989), the details of the water production reactions on the palladium layer are still not known. This phenomenon turns out to be a significant disadvantage of the piezoelectric quartz crystal detector compared to some other solid state gas sensors, e.g., the photopyroelectric device (see Chapter 6).

2.5. FEATURES OF HYDROGEN–SURFACE INTERACTIONS

In this section we review some of the most important conclusions put forward by Lundström et al. (1989) concerning the hydrogen–surface interactions.

1. The steady-state response to hydrogen (low pressures) in any carrier gas follows a relation of the form:

$$\delta\xi_s = \delta\xi_{max}\left\{\frac{\alpha[p(H_2)]^{1/2}}{1 + \alpha[p(H_2)]^{1/2}}\right\} \qquad (2.20)$$

where (a) in an inert atmosphere, $\alpha \equiv K(T)$ (see Eq. 2.13) and (b) in air (or pure oxygen), where $p(H_2) \ll p(O_2)$:

$$\alpha \propto \left[\frac{1}{p(O_2)}\right]^{1/2} \qquad \text{(Case 1)} \qquad (2.21)$$

$$\alpha \propto \left[\frac{1}{p(O_2)}\right]^{1/4} \qquad \text{(Case 2)} \qquad (2.22)$$

2. The sensitivity to hydrogen is several orders of magnitude greater in the presence of an inert atmosphere than in the presence of oxygen (Lundström et al., 1985, 1989).

3. According to Lundström et al. (1989), "in oxygen, Θ_i is independent of the heat of adsorption of hydrogen both at the surface and at the interface. In oxygen, Θ_i, is therefore, not determined by a thermodynamic equilibrium but by steady-state chemical reactions on the metal surface."

4. Detection above 50 °C has the advantage of preventing the adsorption of water molecules and plays a positive role to the speed of response.

5. The history of the catalytic surface plays a major role to the detection abilities of the hydrogen sensing device (sensitivity, selectivity, and time response). In fact, if kept for several months free from the ambient air, the exposure of the active surfaces to H_2 initially exhibits a very slow response but can be reactivated after several exposures (Lundström, 1981; Christofides and Mandelis, 1989a,b; Christofides et al., 1991).

6. The sticking coefficient for a clean palladium surface in UHV (ultrahigh vacuum) is close to 1 (Petersson et al., 1985) and as low as 10^{-4} to 10^{-6} under STP (standard temperature and pressure) conditions in an inert atmosphere.

2.6. WHY HYDROGEN DETECTION?

In recent years, hydrogen has grown to be one of the most useful gases. In many fields such as the chemical, food, metallurgical, electronic, and nuclear industries, hydrogen is increasingly assuming the role of a raw material (Vaz de Campos, 1987). Some examples of the importance of detecting hydrogen gas are as follows: (a) H_2 is absorbed by several metals, causing the precipitation of hydrides in titanium and zirconium and, in the case of steel, hydrogen embrittlement. (b) Most pipework in a chemical plant is encased in insulation in order to prevent heat loss; however, if water penetrates, induced undetected corrosion occurs. (c) During electroplating, hydrogen often evolves and enters the metal substrate. Since the oil crisis of 1973 hydrogen gas has attracted public attention as a clean energy resource, because of its high heat capacity. One reason that hydrogen may play an important energy role in the near future is that use of this gas may enable the world to avoid a mean global warming of at least $2\,°C$ by the mid-twenty-first century caused by the greenhouse effect. Furthermore, it is because of its anticipated usefulness that hydrogen is expected to play a role of increasing importance primarily as an intermediary in the manufacture of synthetic fuels (Dell, 1982).

Research and development work in advanced on-board fuel is of foremost importance for the transportation industry. Specialists believe that hydrogen-fueled motor vehicles will soon be a reality (Peschka, 1987). However, the increasing use of hydrogen gas in not without some disadvantages. In fact, storage of this gas poses a number of problems. A hydrogen leak in large quantities should be avoided because hydrogen when mixed with air at or above 4.65 to 93.9 vol.% is explosive (Weast, 1976). For this reason, among others, it has become very important to develop highly sensitive hydrogen detectors to prevent accidents due to H_2 gas leakage. The Appendix (see p. 313) presents a useful comparison of operating characteristics of several types of hydrogen gas sensors. Such detectors should allow continuous monitoring of the concentration of gases in the environment in a quantitative and selective way. In addition, this control system should be economical.

Another area in which hydrogen gas detectors are very useful is the field of surface science. In fact, devices operating on the principle of a decrease of the work function of Pd by adsorption of H_2 are the main means for the study of surface–gas interaction in the Pd–H_2 system.

REFERENCES

Abe, A., and Hosoya, T. (1984). *Proc. World Hydrogen Energy Conf.*, 5th, Toronto, 1984, Vol. 4, p. 1893.

Armgarth, M., Söderberg, D., and Lundström, I. (1982). *Appl. Phys. Lett.* **41**, 654.

Banerjee, S., and Lee, M.H. (1979). *J. Appl. Phys.* **50**, 1776.

Behm, R.J., Christmann, K., and Ertl, G. (1980). *Surf. Sci.* **99**, 320.

Bucur, R.V., and Mecea, V. (1980). *Surf. Technol.* **11**, 305.

Bucur, R.V., Mecea, V., and Indrea, E. (1976). *J. Less-Common Met.* **49**, 147.

Christofides, C., and Mandelis, A. (1989a). *J. Appl. Phys.* **66**, 3975.

Christofides, C., and Mandelis, A. (1989b). *J. Appl. Phys.* **66**, 3986.

Christofides, C., Mandelis, A., and Enright, J. (1991). *Jpn. J. Appl. Phys.* **30**, 2916.

Conrad, H., Ertl, G., and Latta, E.E. (1974). *Surf. Sci.* **41**, 435.

Daniels, F., and Alberty, R.A. (1961). *Physical Chemistry*, 2nd ed. Wiley, New York.

Davenport, J.W., Diences, G.J., and Johnson, R.A. (1982). *Phys. Rev. B* **25**, 2165.

Dell, R.M. (1982). In *Solid-State Protonic Conductors for Fuel Cells and Sensors*. (J.B. Goodenough, J. Jensen, and M. Kleitz, Eds.), Part II, p. 13. Odense University Press, Odense, Denmark.

Engel, T., and Kuipers, H. (1979). *Surf. Sci.* **90**, 162.

Gentry, S.J., and Jones, T.A. (1986). *Sens. Actuators* **10**, 141.

Gomer, R. (1953). *J. Phys. Chem.* **21**, 1869.

Gysae, B., and Wagener, S. (1940). *Z. Phys.* **115**, 296.

Hong, K.C., and Sapru, K. (1987). *Int. J. Hydrogen Energy* **12**, 165.

Keramati, B., and Zemel, J.N. (1978). *Phys. SiO_2 Its Interfaces, Proc. Int. Top. Conf.*, Yorktown Heights, NY, 1978, p. 459.

Kohl, D. (1990). *Sens. Actuators* **B1**, 158.

Lewis, F.A. (1967). *The Palladium–Hydrogen System*. Academic Press, New York.

Lundström, I. (1981). *Sen. Actuators* **1**, 403.

Lundström, I., Shivaraman, M.S., and Svensson, C. (1985). *Surf. Sci.* **64**, 497.

Lundström, I., Armgarth, M., and Petersson, L.G. (1989). *CRC Crit. Rev. Solid State Mater. Sci.* **15**, 201.

Macly, G.J. (1985). *IEEE Trans. Electron Devices* **ED-32**, 1158.

Mignolet, J.C.P. (1950). *Discuss. Faraday Soc.* **8**, 326.

Nakatsuji, H., Hada, M., and Yonezawa, T. (1987). *J. Am. Chem. Soc.* **109**, 1902.

Peschka, W. (1987). *Int. J. Hydrogen Energy* **12**, 753.

Petersson, L.G., Dannetun, H.M., and Lundström, I. (1985). *Surf. Sci.* **161**, 77.

Picard, C., Kleppa, O.J., and Boureau, G. (1978). *J. Chem. Phys.* **69**, 5549.

Ponec, V., Knor, Z., and Cerny, S. (1969). *Discuss. Faraday Soc.* **41**, 149.

Qian, S., and Northwood, D.O. (1988). *Int. J. Hydrogen Energy* **13**, 25.

Riviere, J.C. (1970). *Solid State Surf. Sci.* **2**, 118.

Simons, J.W., and Flanagan, T.B. (1965). *J. Phys. Chem.* **69**, 3773.

Somorjai, G.A. (1972). *Principles of Surface Chemistry*, Chapter 5. Prentice-Hall, Englewood Cliffs, NJ.

Thomas, J.M., and Thomas, W.J. (1967). *Introduction to the Principles of Heterogeneous Catalysis*, Chapter 2. Academic Press, London.

Vannice, M.A., Benson, J.E., and Boudart, M. (1970). *J. Catal.* **16**, 348.

Vaz de Campos, E.F.P. (1987). *Int. J. Hydrogen Energy* **12**, 847.

Weast, R.C. (1976). *Handbook of Chemistry and Physics*, p. D-107. CRC Press, Cleveland, OH.

CHAPTER

3

GAS-SENSITIVE SOLID STATE SEMICONDUCTOR SENSORS

3.1. INTRODUCTION

The field of solid state gas sensors based on semiconductor devices encompasses junction structures in the general categories of metal–semiconductor (Schottky barrier) diodes, metal–insulator–semiconductor (MIS) capacitors, metal–oxide–semiconductor field effect transistors (MOSFETs), and thick/thin surface film oxide semiconductor devices. In the case of gas-sensitive Schottky barrier diodes, their fabrication consists of depositing a catalytic metal layer on a clean or very thinly oxidized semiconducting substrate. For example, hydrogen-gas-sensitive devices can be fabricated using Pd–Si, Pd–silicide–Si, and Pd–silicon dioxide–Si geometries, with insulator thicknesses small enough to allow carrier (primarily electron) tunneling across the interface layer. MOS devices very sensitive to small gas concentrations can be fabricated by oxidizing single-crystalline n-type or p-type Si substrates in controlled atmospheres of oxygen or water vapor, in order to grow a desirable thickness of SiO_2 on the surface. A thin metal gate is subsequently deposited on the oxide to form the biasing electrode, which is made of a catalytic metal in the specific case of gas-sensing devices. The type of sensors fabricated in this manner can be either a MOS capacitor or a transistor. The capacitor structure is simpler and usually consists of metal and gate-oxide dielectric thin layers (both *ca.* 1000 Å thick) on n- or p-type Si substrate, the resistivity of which is a few ohm-centimeters. The MOS transistor structure requires some additional processing steps: the formation of n-type contacts in the silicon through diffusion or ion implantation in specific regions on either end of the oxide layer with a channel distance between them on the order of 100 Å. Cross-sectional views of the basic MOS capacitor and transistor structures are shown in Fig. 3.1.

In the category of thick-film semiconducting surface devices one can include gas-sensitive resistors, such as ZnO and SnO_2. It has been known for more than a quarter century that the resistivity of such semiconductors is strongly dependent on the ambient concentration of hydrocarbon vapors (Heiland, 1982).

19

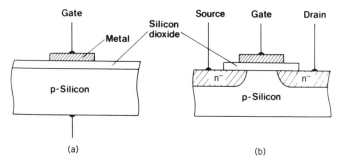

Figure 3.1. Cross-sectional views of (a) the basic MOS capacitor and (b) the MOS transistor (Lundström and Svensson, 1985).

Thin organic semiconductor films (e.g., phthalocyanines) have also exhibited changes in their resistivity upon exposure to certain gases, such as ammonia (Wohltjen et al., 1985).

Other important non-FET thin-film structures fabricated using semi-conductor heterojunctions have been based on controlling the surface potential of the substrate semiconductor through the gas uptake by the surface heterostructure (Lindmeyer, 1965; Zemel, 1975). Such devices include the PbS–Si and Pd–CdS diode structures, which have exhibited good sensitivities to hydrogen uptake (Zemel et al., 1975; Steele and MacIver, 1976).

In this chapter we shall present the basic physics of the operation of the four classes of semiconductor devices mentioned above and we shall give examples of their applications in the gas sensor field through a critical selection of the considerable literature which has accumulated in the last 25 years. In order to keep the extent of this chapter within reasonable limits, it is assumed that the reader is familiar with elementary semiconductor physics (Kittel, 1976; Madelung, 1978). A concise and lucid review of the physical aspects of solid state theory, which are pertinent to the fabrication and operation of semiconductor gas sensors can be found in Zemel (1981). Details of the physics of the devices presented here are given in Size (1981).

3.2. METAL–SEMICONDUCTOR (SCHOTTKY BARRIER) DEVICES

Under equilibrium conditions, when a metal is brought into intimate contact with a semiconductor the Fermi levels on both sides of the junction are aligned: the Fermi level of the semiconductor adjusts itself to the dominating presence of the metal (the free carrier density of which is much higher than that of the semiconductor) and is lowered by an amount equal to the difference between the two work functions. This adjustment of Fermi levels occurs even if there exists a thin dielectric layer, such as SiO_2, between the metal and the semiconductor. The situation is shown in Fig. 3.2 (Henisch, 1957), which depicts the energy band diagram of the Schottky barrier structure in the absence of surface states in the semiconductor. A barrier height equal to (Sze, 1981)

$$q\phi_{Bn} = q(\phi_m - \chi) \tag{3.1}$$

is typical of the metal–n-type semiconductor junction, as shown in Fig. 3.2(a-4). Equation (3.1) is approximately valid, once the potential lowering

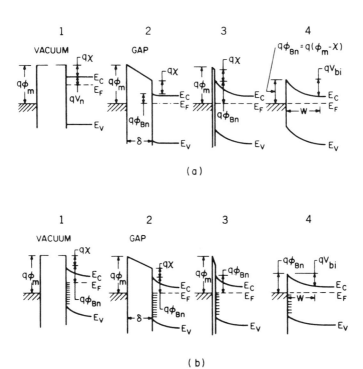

(a)

(b)

Figure 3.2. (a) Energy-band diagrams of metal–oxide–n-type semiconductor junction: (1) before contact; (2) after contact and thermodynamic equilibrium is attained, in the presence of a gap (or an insulating dielectric) of thickness δ: (3) when δ is comparable to interatomic distances; and (4) in the limit $\delta = 0$ (metal–semiconductor junction). Here $q\phi_m$ is the metal work function: $q\chi$, electron affinity in the semiconductor; qV_n equals $E_c - E_F$, where $E_c(E_F)$ is the energy at the bottom of the conduction band (Fermi energy or Fermi level); E_v, energy at the top of the valence band; $q\phi_{Bn}$, barrier height; V_{bi}: built-in potential (Henisch, 1957); (b) Same as in (a) but in the presence of surface (interface) states in the semiconductor.

due to the Schottky effect (Sze, 1981) is neglected. If the barrier height of Eq. (3.1) is high enough, then the Schottky diode is rectifying, i.e., it exhibits asymmetric current–voltage $(I-V)$ characteristics, which are highly nonlinear. Upon negatively biasing the metal–thin oxide–n-type semiconductor structure of Fig. 3.2(a-3), beyond the depletion configurations, there may result a strong enough upward energy band bending in the n-type semiconductor, so that the negative bias in the metal inverts the semiconductor surface by increasing the interfacial hole population, thereby converting it to the p-type. In this case, the diode behaves like an ordinary p-n junction in series with a thin tunneling layer. Upon exposure to certain gases the Schottky barrier height is affected. The dependence of $q\phi_{Bn}$ on the metal work function, $q\phi_m$, indicates that gas-sensitive devices can result when Schottky diodes are exposed to gases that alter the metal work function. The bending of the semiconductor bands at the interface occurs down to a depth W (Fig. 3.2), which is called the depletion width or the space-charge layer. Under the abrupt approximation, the carrier density (electrons in n-type semiconductors) is given by

$$\rho(x) \simeq \begin{cases} qN_D; & x < W \\ 0 \quad ; & x > W \end{cases} \qquad (3.2)$$

where N_D is the donor density. Furthermore, the electric field $-dV(x)/dx \simeq 0$ for $x > W$.

Solution to Poisson's equation

$$\frac{d^2 V(x)}{dx^2} = -\frac{1}{\epsilon_s}\rho(x) \qquad (3.3)$$

gives the extent of the depletion width in the presence of a bias V:

$$W(V) = \left[\frac{2\epsilon_s}{qN_0}\left(V_{bi} - V - \frac{kT}{q} \right) \right]^{1/2} \qquad (3.4)$$

where T is the absolute temperature; ϵ_s, the permittivity of the semiconductor; and V_{bi}, the built-in potential [Fig. 3.2(a-4, b-4)]. The same equation gives the value of the space charge Q_{sc} per unit area of the semiconductor

$$Q_{sc}(V) = qN_d W = \left[2q\epsilon_s N_D\left(V_{bi} - V - \frac{kT}{q} \right) \right]^{1/2} \quad [C/cm^2] \quad (3.5)$$

Now Eq. (3.5) can be used to determine the depletion layer differential capacitance per unit area, a key experimental quantity:

$$C(V) \equiv \frac{|\partial Q_{sc}(V)|}{\partial V} = \left[\frac{q\epsilon_s N_D}{2\left(V_{bi} - V - \dfrac{kT}{q} \right)} \right]^{1/2} = \frac{\epsilon_s}{W} \qquad [\text{F/cm}^2] \qquad (3.6)$$

The foregoing considerations are ideal and valid only for strongly ionic semiconductors and not for covalent materials such as silicon (Kurtin et al., 1969). In the latter case significant surface state densities are present at the metal–semiconductor interface and it is these states that control the barrier height, $q\phi_B$, rather than the metal work function, as indicated in Eq. (3.1). The appearance of surface states is the result of the discontinuity in the band structure of the semiconductor imposed by the presence of the surface, and they are independent of the metal on the other side of the junction. Figure 3.3 shows the Schottky diode energy diagram for an n-type semiconductor with surface states present (Cowley and Sze, 1965). The various parameters used to describe the metal–semiconductor barrier are defined in the figure. In this realistic approach to a Schottky barrier a density D_s of acceptor surface states is assumed, D_s (states/cm$^2 \cdot$eV) being constant over the energy range between the surface energy level, $q\phi_0$, and the Fermi level. In Fig. 3.3 the surface barrier lowering, $q\,\Delta\phi$, due to the Schottky effect has also been taken into account, and the surface state and space-charge densities at the thermal equilibrium are given by

$$Q_{ss} = -qD_s(E_g - q\phi_0 - q\phi_{Bn} - q\Delta\phi) \qquad [\text{C/cm}^2] \qquad (3.7)$$

and

$$Q_{sc} = \left[2q\epsilon_s N_D \left(\phi_{Bn} - V_n + \Delta\phi - \frac{kT}{q} \right) \right]^{1/2} \qquad [\text{C/cm}^2] \qquad (3.8)$$

where Eq. (3.8) differs from Eq. (3.5) only in its inclusion of the depletion layer decrease due to the Schottky effect under forward biasing conditions (Rideout, 1978). Of importance to mechanisms leading to Schottky diode sensitivity to certain ambient gas concentrations is the current transport process of the junction. This process consists of four basic contributions under forward bias, with the inverse processes occurring under reverse bias (Rhoderick, 1978): (1) electron transport from the semiconductor over the potential barrier (Fig. 3.4) into the metal (this is the dominant process for Schottky diodes with moderately doped semiconductors operating at or near room temperature); (2) quantum mechanical tunneling of electrons through the barrier in heavily doped semiconductors; (3) electron-hole recombination

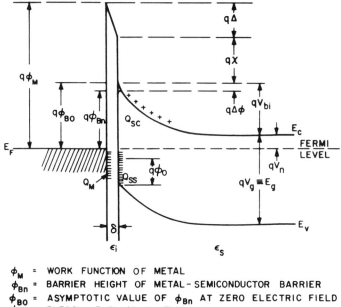

ϕ_M = WORK FUNCTION OF METAL
ϕ_{Bn} = BARRIER HEIGHT OF METAL - SEMICONDUCTOR BARRIER
ϕ_{BO} = ASYMPTOTIC VALUE OF ϕ_{Bn} AT ZERO ELECTRIC FIELD
ϕ_0 = ENERGY LEVEL AT SURFACE
$\Delta\phi$ = IMAGE FORCE BARRIER LOWERING
Δ = POTENTIAL ACROSS INTERFACIAL LAYER
χ = ELECTRON AFFINITY OF SEMICONDUCTOR
V_{bi} = BUILT - IN POTENTIAL
ϵ_s = PERMITTIVITY OF SEMICONDUCTOR
ϵ_i = PERMITTIVITY OF INTERFACIAL LAYER
δ = THICKNESS OF INTERFACIAL LAYER
Q_{sc} = SPACE - CHARGE DENSITY IN SEMICONDUCTOR
Q_{ss} = SURFACE - STATE DENSITY ON SEMICONDUCTOR
Q_M = SURFACE - CHARGE DENSITY ON METAL

Figure 3.3. Energy-band diagram of a metal–n-type semiconductor contact with an interfacial layer of the order of an atomic distance (Cowley and Sze, 1965).

in the space-charge region; and (4) hole injection from the metal into the semiconductor, leading to recombination in the neutral n-region. The electron transport over the potential barrier has been discussed in terms of thermionic emission diffusion theory by Crowell and Sze (1966), who have shown that the complete expression of the current density–voltage (J–V) characteristics for the Schottky diode is

$$J(V) = J_s[\exp(qV/kT) - 1] \tag{3.9}$$

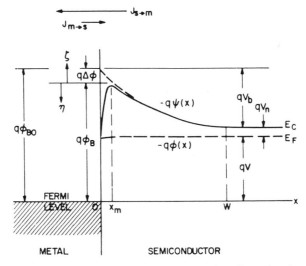

Figure 3.4. Energy-band diagram incorporating the Schottky effect. The electron potential energy is $q\psi(x)$, and the quasi-Fermi level of the n-type semiconductor is $q\phi(x)$. The rest of the quantities are similar to those defined in Fig. 3.3 (Sze, 1981).

and the saturation current density

$$J_s = A^{**}T^2 \exp(-q\phi_{Bn}/kT), \tag{3.10}$$

where A^{**} is the effective Richardson constant (A/cm$^2 \cdot$K^2) given by

$$A^{**} = \frac{f_p f_Q A^*}{1 + f_p f_Q (v_R/v_D)} \tag{3.11}$$

In Eq. (3.11) A^* is a simpler "effective Richardson constant":

$$A^* = 4\pi q m^* k^2/h^3 \tag{3.12}$$

which is obtained neglecting the effects of optical phonon scattering and quantum mechanical reflections at the metal–semiconductor interface; m^* is the electronic effective mass in the semiconductor. For free electrons $A^* \equiv A = 120 \, \text{A/cm}^2 \cdot \text{K}^2$. In the more complete version of Crowell and Sze (1966), f_p is the probability of electron emission over the potential maximum x_m (Fig. 3.4) and is given by $f_p = \exp(-x_m/\lambda)$, where λ is the carrier mean free path. The function f_Q is the ratio of the total current flow when the quantum mechanical tunneling and reflection are taken into account, to the current flow when these effects are neglected; v_R is the carrier thermal velocity

$$v_R = A^* T^2 / q N_c \qquad (3.13)$$

where N_c is the effective density of states in the conduction band; v_D is an effective diffusion velocity, such that if $v_D \ll v_R$ the diffusion process across the barrier is dominant; if $v_D \gg v_R$, the simple thermionic emission theory (Bethe, 1942) is applicable with $A^{**} \to A^*$.

For moderately doped semiconductors the other three transport processes are relatively unimportant and the $J-V$ characteristics of the Schottky diode are well described by Eq. (3.9). Figure 3.5 shows two typical curves of W–Si and W–GaAs diodes biased in the forward direction (Crowell et al., 1965):

$$J(V) \simeq A^{**} T^2 \exp\left(-\frac{q\phi_{B0}}{kT} \right) \exp\left[\frac{q(\Delta\phi + V)}{kT} \right] \qquad (3.14)$$

where ϕ_{B0} is the zero-field asymptotic barrier height (Fig. 3.3), and $\Delta\phi$ is the Schottky barrier lowering (image-force potential). For $V > 3kT/q$ the forward current density can be expressed as

$$J \simeq J_s \exp(qV/nkT) \qquad (3.15)$$

where n is the ideality factor: $n = 1.02$ for the W–Si diode, and $n = 1.04$ for the W–GaAs diode. The saturation current density, J_s, is the extrapolated current density value at $V_F = 0$. The barrier height can then be obtained by using Eq. (3.10):

$$\phi_{Bn} = \frac{kT}{q} \ln\left(\frac{A^{**} T^2}{J_s} \right) \qquad (3.10a)$$

with $A^{**} = 120 \, \text{A/cm}^2 \cdot \text{K}^2$. The value of ϕ_{Bn} is not very sensitive to the choice of A^{**}. The barrier height can also be determined by measuring the differential capacitance of the Schottky diode, as given in Eq. (3.6). That equation indicates that plots of $1/C^2$ vs. applied bias V should be straight lines. Goodman (1963) and Crowell et al. (1965) have shown that the barrier height can then be determined from the intercept on the voltage axis:

$$\phi_{Bn} = V_i + V_n + \frac{kT}{q} - \Delta\phi \qquad (3.16)$$

where V_i is the voltage intercept, and V_n is the depth of the Fermi level below

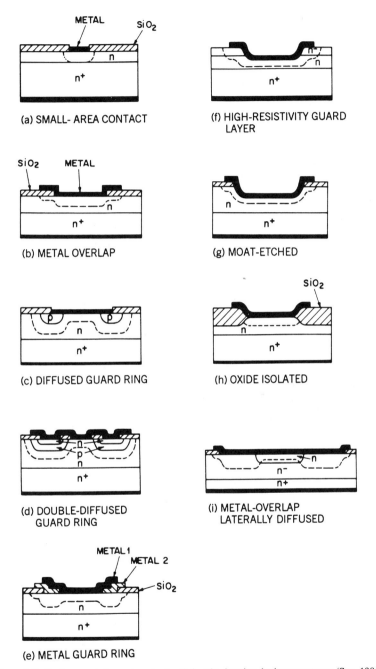

Figure 3.5. Various metal–semiconductor Schottky barrier device structures (Sze, 1981).

the conduction bandedge. This quantity may be computed if the semiconductor doping density is known. The slope of the $1/C^2$ vs. V line can also be used to extract the carrier doping density.

Finally, Fig. 3.5 shows various metal–semiconductor device structures as options for gas sensor device fabrication, many of which have already been employed toward this end. Figure 3.5(a) is a small-area contact device fabricated by planar processing on epitaxial n on n^+ substrate (Ryder, 1968). Typical applications as a minimum series resistance and capacitance diodes (i.e., short response time constant) have been implemented using Au–Si and Au–GaAs structures. The metal overlap structure [Fig. 3.5(b)], gives nearly ideal J–V characteristics and low leakage current at moderate reverse bias, with increasing reverse current at large reverse biases, owing to the electrode sharp-edge effect. To eliminate this effect, Fig. 3.5(c) uses a diffused guard ring to give nearly ideal forward *and* reverse characteristics (Fig. 3.6). This structure suffers from long recovery time and large parasitic capacitance due to the neighboring p–n junction. Figure 3.5(d) uses a double-diffused guard ring to reduce the recovery time, but the process is relatively complicated. Figure 3.5(e) suggests using two metals with different barrier heights, large variations in which are a difficult realization with covalent semiconductors. Figure 3.5(f) shows a Schottky barrier diode exhibiting a high resistivity layer on top of the active layer, with a parasitic capacitance generally higher than that of Fig. 3.5(b). A different approach consists of surrounding the diode by a moat [Fig. 3.5(g)], which may lead to problems from burying contaminants in the moat; or one can use oxide isolation [Fig. 3.5(h)] to reduce the sharp-edge field. This latter configuration requires a special planar process in performing the local oxidation. Figure 3.5(i) is a metal-overlap laterally diffused structure, basically a double parallel Schottky diode giving nearly ideal forward and reverse J–V characteristics and a very short reverse recovery time. These features are achieved at the expense of some more complicated processing than other Schottky diodes.

3.3. METAL–INSULATOR–SEMICONDUCTOR (MIS) DEVICES

3.3.1. The MIS Capacitor

A detailed and comprehensive treatment of this very important class of devices can be found in the book by Nicollian and Brews (1982). The primary importance of MIS devices in the gas sensor field lies in the sensitivity of their voltage response to the surface conditions. The single most important MIS subclass is the metal–oxide–semiconductor (MOS) device family, with metallized SiO_2–Si being the most widely used structure in both electronic as well

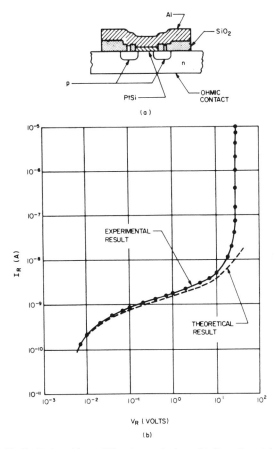

Figure 3.6. (a) PtSi–Si–diode with a diffused guard ring. (b) Experimental and theoretical near-ideal forward and reverse I–V characteristics (Lepselter and Sze, 1968).

as gas sensor fabrication. Figure 3.1(a) shows the cross-sectional view of the basic MIS diode. Figure 3.7 describes the energy-band diagrams for p-type and n-type ideal MIS diodes: E_F indicates the local Fermi level in the metal and the semiconductor, and E_i is the intrinsic Fermi level. The thickness of the insulator is d and V is the applied voltage on the metal. For an ideal diode, the work function difference between metal and semiconductor is zero at zero applied bias, i.e., the semiconductor energy bands are flat (flat-band condition). Under all biasing conditions it is also assumed that no charges can be found in the insulator: two equal charges of opposite sign exist on the metal and semiconductor side of the interlayer. No carrier transport through the insulator is assumed to exist under dc bias. In Fig. 3.7 three

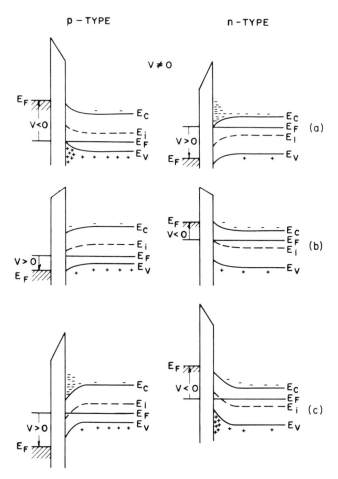

Figure 3.7. Energy-band diagrams for ideal MIS diodes under nonzero applied bias ($V \neq 0$), corresponding to the cases: (a) accumulation; (b) depletion; (c) inversion (Sze, 1981).

cases of interest are described: For a p-type semiconductor the application of a negative voltage, $V < 0$, to the metal results in the upward bending of the bands due to an accumulation of majority carriers (holes) near the semiconductor surface caused by the attractive ($V < 0$) potential field. No current flows; therefore the position of E_F remains constant. If $V > 0$, then the semiconductor bands bend downward, a result of the repulsive potential that drives majority carriers (holes) away from the semiconductor surface, thus depleting the region of free carriers. For metal voltage values $V \gg 0$, steeper bending of the bands occurs, so that $E_i < E_F$ near the surface of the semiconductor. A strong electric field attractive to minority carriers (electrons) exists, and the condition $E_i < E_F$ results in the minority carrier number density becoming larger than the majority number density at and near the surface (inversion). The hole concentration has been depleted so severely that the electron concentration dominates according to the mass action law (Kittel, 1976):

$$n(T)p(T) = n_i^2(T) \qquad (3.17)$$

where $n(T)$ and $p(T)$ are the number densities of free electrons and holes, repectively, at absolute temperature T; and $n_i(T)$ is the intrinsic carrier density for the same semiconductor at the same temperature. Similar results (opposite polarity) can be obtained for the n-type semiconductor in Fig. 3.7.

In the case where a MIS structure operates as a capacitor, the capacitance of the semiconductor depletion layer is an important parameter of central interest to gas sensor device characteristics. Assuming a p-type semiconductor substrate, the detailed configuration of the energy-band diagram is shown in Fig. 3.8. The potential $\psi(x)$ can be obtained from the one-dimensional Poisson equation, Eq. (3.3), with the total space charge density $\rho(x)$ given

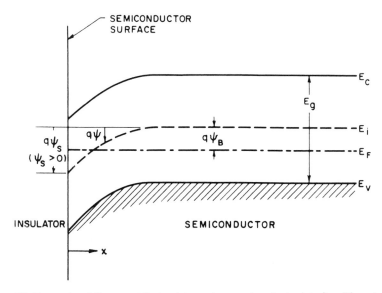

Figure 3.8. Energy-band diagram at the insulator–p-type semiconductor interface. The potential ψ, defined as zero in the bulk, is measured with respect to the intrinsic Fermi level, E_i. The surface potential ψ_s is positive as shown. Accumulation occurs when $\psi_s < 0$. Depletion occurs when $\psi_B > \psi_s > 0$. Inversion occurs when $\psi_s > \psi_B$ (Sze, 1981).

by (Sze, 1981):

$$\rho(x) = q(N_D^+ - N_A^- + p_p - n_p) \tag{3.18}$$

where N_D^+ (N_A^-) is the density of ionized donors (acceptors) and p_p (n_p) is the hole (electron) concentration in the semiconductor, a function of $\psi(x)$:

$$p_p(x) = p_{po} \exp[-q\psi(x)/kT] \tag{3.19a}$$

$$n_p(x) = n_{po} \exp[q\psi(x)/kT] \tag{3.19b}$$

p_{po} and n_{po} are the equilibrium hole and electron densities, respectively, in the bulk of the semiconductor. It should be remembered that ψ is positive when the bands are bent downward, as in Fig. 3.8. Charge neutrality in the semiconductor bulk dictates $\lim_{x \to \infty} \rho(x) = \lim_{x \to \infty} \psi(x) = 0$, or

$$N_D^+ - N_A^- = n_{po} - p_{po} \tag{3.20}$$

Once integrating Poisson's equation from the bulk to the surface yields the relation between the electric field $[\mathcal{E}(z) = -d\psi(x)/dx]$ and the potential ψ:

$$\mathcal{E}(x) = \pm \frac{\sqrt{2kT}}{qL_D} F\left[\beta\psi(x), \frac{n_{po}}{p_{po}} \right] \tag{3.21}$$

where L_D is the extrinsic Debye length for holes:

$$L_D \equiv [\mathcal{E}_s / q p_{po} \beta]^{1/2}; \qquad \beta \equiv q/kT \tag{3.22}$$

and

$$F\left(\beta\psi, \frac{n_{po}}{p_{po}} \right) \equiv \left[(e^{-\beta\psi} + \beta\psi - 1) + \frac{n_{po}}{p_{po}}(e^{\beta\psi} - \beta\psi - 1) \right]^{1/2} \geqslant 0 \tag{3.23}$$

In Eq. (3.21) the positive (negative) sign corresponds to the case $\psi > 0$ $(\psi < 0)$. The space charge per unit area required to produce the surface electric field, $\mathcal{E}_s = E(0)$, can be determined using Gauss's law (Wangsness, 1986):

$$Q_s = -\epsilon_s \mathcal{E}_s = \mp \frac{\sqrt{2kT\epsilon_s}}{qL_D} F\left(\beta\psi_s, \frac{n_{po}}{p_{po}} \right) \tag{3.24}$$

Finally, the differential capacitance of the semiconductor depletion layer can be calculated from Eqs. (3.23) and (3.24):

$$C_D \equiv \frac{\partial Q_s}{\partial \psi_s} = \frac{\epsilon_s}{\sqrt{2}L_D} \frac{[1 - e^{-\beta\psi_s} + (n_{po}/p_{po})(e^{\beta\psi_s} - 1)]}{F(\beta\psi_s, n_{po}/p_{po})} \qquad [\text{F/cm}^2] \quad (3.25)$$

Figure 3.8 can be used to show that $\psi_s = 0$ at flat bands, in which case C_D is considerably simplified:

$$C_D|_{\text{flat band}} = \epsilon_s/L_D \qquad [\text{F/cm}^2] \tag{3.26}$$

For ideal MIS capacitors, i.e., in the absence of work function differences between metal and semiconductor, the total capacitance C of the system is a series combination of the insulator capacitance

$$C_i = \epsilon_i/d \tag{3.27}$$

and the semiconductor depletion-layer capacitance C_D:

$$C(V) = \frac{C_i C_D(V)}{C_i + C_D(V)} \qquad [\text{F/cm}^2] \tag{3.28}$$

Figure 3.9 gives the behavior of the complete MIS structure capacitance as a function of the applied voltage V, such that

$$V = \psi_s + V_i \tag{3.29}$$

where V_i is the potential drop across the insulator

$$V_i = \epsilon_i d = |Q_s|/C_i \tag{3.30}$$

The behavior of $C(V)$ vs. V exhibited in Fig. 3.9 can be explained by starting at the left-hand side ($V < 0$), corresponding to accumulation of holes for the p-type material. The effective charge thickness in the semiconductor is very small, which, in turn, renders $C_D \gg C_i$. Equation (3.28) shows that $C(V) \simeq C_i$, which is a constant [Eq. (3.27)], independent of applied voltage. As the negative voltage decreases, the effective charge (holes) thickness increases as a depletion layer is formed that acts as a dielectric in series with the insulator and $C(V)$ decreases. As the voltage V changes sign, negative charges (electrons) are attracted to the semiconductor surface, which results in a small effective charge thickness (of opposite polarity) in the semiconductor, giving $C_D \gg C_i$, once again, for high positive V (strong inversion). This mechanism tends to

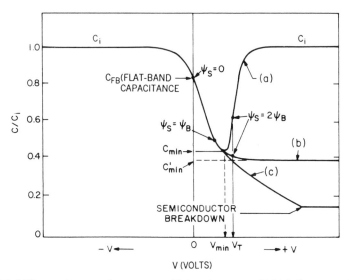

Figure 3.9. MIS capacitance–voltage curves: (a) low frequency; (b) high frequency; (c) deep depletion (Grove et al., 1965).

restore the total capacitance to its $V \ll 0$ value, i.e., $C(V \gg 0) \rightarrow C(V \ll 0) = C_i$. The observed minimum capacitance and the corresponding minimum voltage are designated C_{min} and V_{min}, respectively, in Fig. 3.9, curve (a).

The capacitive behavior of the MIS diode is also a function of the modulation frequency of the applied (ac) voltage, since carrier contributions to the capacitance depend on their ability to follow the applied ac field. This is only feasible at low frequencies, such that the generation and recombination rates of minority carriers can follow the ac signal variation and lead to synchronous charge exchange with the inversion layer. In the metal–SiO$_2$–Si system the critical frequency above which no synchronous contribution of minority carriers with the applied voltage can be observed lies between 5 and 100 Hz (Grove et al., 1965). At higher frequencies than minority carrier generation rates, the inversion layer is not capable of fully developing and the capacitance exhibits a small increase or none in the $V > 0$ range (Figs. 3.9 and 3.10). The minimum value of $C(V)$ in Fig. 3.10 has been shown to be (Brews, 1977)

$$C'_{min} \simeq \frac{\epsilon_i}{d + (\epsilon_i/\epsilon_s)W_m} \tag{3.31}$$

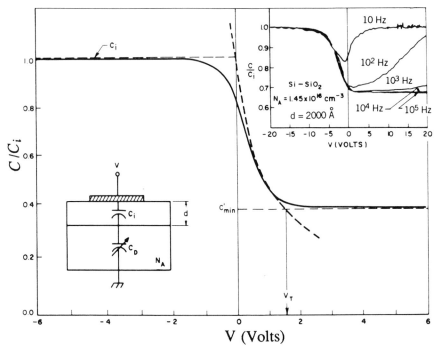

Figure 3.10. High-frequency metal–SiO_2–Si capacitance–voltage curve showing its approximate segments (dashed lines). The insert shows experimental results of the frequency effect (Grove et al., 1965).

where

$$W_m \simeq \frac{2}{q}\left[\frac{\epsilon_s kT \ln(N_A/n_i)}{N_A}\right]^{1/2} \qquad (3.32)$$

is the maximum width of the semiconductor surface depletion layer under steady state conditions. From the inset of Fig. 3.10 it can be seen that the low-frequency behavior sets in at $f \lesssim 100\,Hz$.

It is important for the understanding of the behavior of MIS-based gas sensor devices to emphasize that real (rather than ideal) structures involving the Si–SiO_2 interface still present challenges in the interpretation of the interfacial electronic phenomena. An appealing picture (Nicollian and Brews, 1982; Sze, 1983) describes the thermally oxidized real interface as a silicon single crystal, followed by a monolayer of SiO_x (incompletely oxidized silicon, $1 < x < 2$), then a strained region of SiO_2 (~ 10–$40\,\text{Å}$ deep), and the remainder a stoichiometric ($x = 2$), strain-free, amorphous oxide. In practical

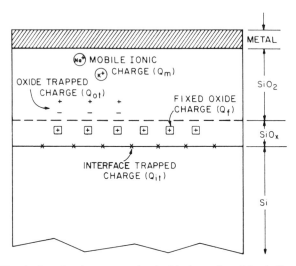

Figure 3.11. Terminology for charges associated with thermally oxidized silicon (Deal, 1980).

MOS diodes, interface traps and oxide charges are known to exist and affect the ideal MOS characteristics. Deal (1980) has provided a useful classification scheme of these traps and charges (Fig. 3.11): (1) Interface-trapped charges, Q_{it}, located at the Si–SiO$_x$ interface with energy states in the silicon bandgap—these charges can interact with silicon through fast charge exchanges; Q_{it} can be produced by excess (trivalent) silicon, excess oxygen, and impurities. (2) Fixed oxide charges, Q_f, located at or near the interface—these charges are immobile under an applied electric field. (3) Oxide-trapped charges, Q_{ot}, which can be created by external means, e.g., ionizing radiation of hot-electron injection—these traps are distributed inside the SiO$_2$ layer. (4) Mobile ionic charges, Q_m, e.g., sodium ions—these charges are mobile within the oxide under bias-temperature aging conditions. In fact, capacitance measurements have been used to evaluate certain aspects of the various kinds of trapped charges appearing in Fig. 3.11.

3.3.2. The MOSFET

The metal–oxide–semiconductor field effect transistor (MOSFET) is the most important element in microelectronic integrated circuit technologies and also in semiconductor gas sensor fabrication (Nicollian and Brews, 1982; Sze, 1983; Bergveld, 1985). This device has several other acronyms, including IGFET (insulated-gate field effect transistor), MISFET (metal–insulator–semiconductor field effect transistor), and MOST (metal-oxide-semiconductor

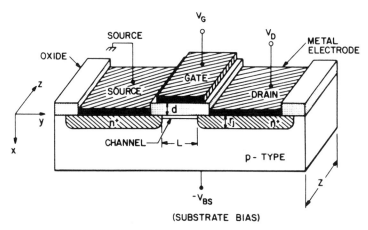

Figure 3.12. Schematic diagram of a MOSFET: V_G, gate voltage; V_D, drain voltage; V_{BS}, substrate bias voltage; d, insulator thickness; L, channel length; Z, channel width; r_j, junction depth (Kahng 1976).

transistor). In conventional MOSFETs current is mainly transported by carriers of one polarity only (unipolar). The most important combination of materials in MOSFET devices is metal(with catalytic properties in the gas sensor case)–SiO_2–Si structures.

The general structural geometry of a MOSFET is illustrated in Fig. 3.12. It is a four-terminal device. With a p-type semiconductor substrate, two ion-implanted n^+ regions form the source and drain. Suitable ions for n^+ implantation are, for instance, P^+ or As^+. The metal deposited on the insulator (usually called "gate oxide") forms the gate. This is an n-channel MOSFET. The channel length L is the distance between the two metallurgical n^+–p junctions. The substrate doping density is N_A. An equivalent p-channel device can also be fabricated. Another way of forming the n^+ regions is through diffusive doping, although this method is less common nowadays. The source contact is grounded and is used as reference. When no bias is applied to the gate ($V_G = 0$), the structure is equivalent to two p–n junctions (source–substrate, substrate–drain) connected back-to-back. The only current that can flow from source to drain is the reverse leakage current due to the different neutral region energy-band levels in the n^+ and p regions. When a sufficiently large positive bias is applied to the gate ($V_G > 0$) to form an inversion layer (or channel) at the semiconductor surface between the two n^+ regions, a large current can flow between the source and the drain, which are now connected by a conducting surface n-channel. The conductance of this channel can be modulated by varying the gate voltage. The substrate bias voltage, V_{BS}, may either be connected to the source or be reverse-biased, and will affect the channel conductance. With the application of a voltage $V_D \neq 0$ across the source-drain contacts the (equal at equilibrium) Fermi levels are disturbed: the hole Fermi level, E_{Fp}, remains at the equilibrium value, whereas the minority carrier (electron) Fermi level is lowered toward the drain contact. Figure 3.13 shows cross-sectional configurations of the

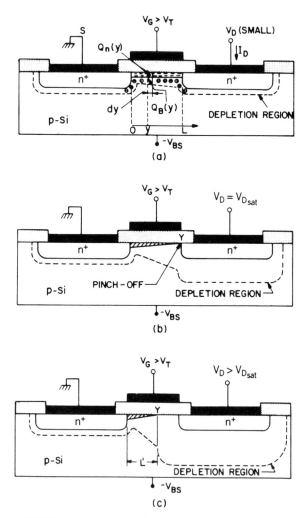

Figure 3.13. (a) MOSFET operated in the linear region (low drain voltage). (b) MOSFET operated at onset of saturation. The point Y indicates the pinch-off point. (c) MOSFET operated beyond saturation with reduced effective channel length (Sze, 1981).

device operation. In Fig. 3.13(a) sufficient voltage is applied to the gate $(V_G > 0)$ to cause inversion at the semiconductor surface. If a small drain voltage V_D is applied, a current will flow from the source to the drain through the conducting n-channel. In this (linear) region the drain current I_D is proportional to the drain voltage. With increasing drain voltage, a value is eventually reached at which the channel depth x_i at $y = L$ is reduced to zero; this is called the pinch-off point [Fig. 3.13(b)]. Beyond the pinch-off point the drain current remains essentially the same, since for $V_D > V_{D\,\text{sat}}$ the voltage at Y remains unaltered, $V_{D\,\text{sat}}$. Therefore, even though the channel length keeps on decreasing [$L' < L$ in Fig. 3.13(c)], the number of carriers arriving at Y from the source remains the same. These carriers are subsequently injected from point Y into the drain depletion region. The ideal MOSFET current–voltage $(I_D$–$V_D)$ characteristics have been derived by Sze (1981) under the following conditions: (1) The metal gate-semiconductor energy-band structure is an ideal MIS Schottky diode as described in Section 3.1; there are no interface traps, fixed oxide charges, work function differences, etc. (2) Diffusion currents are ignored. (3) Carrier mobility in the inversion layer is constant. (4) Doping in the channel is uniform. (5) Reverse leakage current is negligible. (6) The transverse field, $\mathscr{E}_x$, in the channel is much larger than the longitudinal field $\mathscr{E}_y$ (gradual channel approximation). Under these conditions, the total charge per unit area, $Q_s(y)$, induced in the semiconductor at a longitudinal distance y away from the source is given by (Sze, 1981)

$$Q_s(y) = -[V_G + \psi_s(y)]C_i \qquad (3.33)$$

where $C_i = \epsilon_i/d$ is the insulator (or gate) capacitance per unit area, and $\psi_s(y)$ is the semiconductor surface potential. The charge (per unit area) in the inversion layer due to minority carriers is given by

$$Q_n(y) = -[V_G - \psi_s(y)]C_i - Q_B(y) \qquad (3.34)$$

where $Q_B(y)$ is the charge within the inverted surface depletion region W_m:

$$Q_B(y) = -qN_A W_m = -\sqrt{2\epsilon_s qN_A[V(y) + 2\psi_B]} \qquad (3.35)$$

The approximation

$$\psi_s(\text{inv}) \simeq V_D + 2\psi_B \qquad (3.36)$$

was used in Eq. (3.35), where $q\psi_B$ is the energy difference between the Fermi level in the inverted p region and the intrinsic Fermi level E_i in the semiconductor. The channel conductance is given by

$$g = (qZ\mu_n/L) \int_0^{x_i} n(x)\,dx = qZ\mu_n |Q_n(y)|/L \tag{3.37}$$

for a minority carrier density $n(x)$ and constant mobility μ_n; $Q_n(y)$ was defined in Eq. (3.34), and Z is the dimension indicated in Fig. 3.12. Now it can be seen from Fig. 3.13(a) that the channel resistance of an elemental section dy is given by

$$dR = \frac{dy}{gL} = \frac{dy}{qZ\mu_n |Q_n(y)|} \tag{3.38}$$

The voltage drop across this elemental section is

$$dV = I_D\,dR = \frac{I_D\,dy}{Z\mu_n |Q_n(y)|} \tag{3.39}$$

where I_D is the drain current, a constant independent of y. Substituting Eqs. (3.34) and (3.35) into Eq. (3.39) and integrating from the source ($y = 0$, $V = 0$) to the drain ($y = L$, $V = V_D$) yields (Sze, 1981):

$$I_D = \frac{Z}{L}\mu_n C_i \left\{ \left(V_G - 2\psi_B - \frac{V_D}{2} \right) V_D - \frac{2}{3} \frac{\sqrt{2\epsilon_s q N_A}}{C_i} [(V_D + 2\psi_B)^{3/2} - (2\psi_B)^{3/2}] \right\} \tag{3.40}$$

for the ideal case. Equation (3.40) predicts that for a given V_G the drain current first increases linearly with V_D and then gradually levels off, approaching saturation. Figure 3.14 demonstrates this behavior, with the dashed line indicating the locus of the drain voltage ($V_{D\,\text{sat}}$) at which I_D reaches its maximum (saturation, $I_{D\,\text{sat}}$) value. Upon more detailed consideration of Eq. (3.40), for small V_D it reduces to the quadratic form

$$I_D \simeq \frac{Z}{L}\mu_n C_i \left[(V_G - V_T)V_D - \left(\frac{1}{2} + \frac{\sqrt{\epsilon_s q N_A/\psi_B}}{4C_i} \right) V_D^2 \right] \tag{3.41}$$

or, for $V_D \ll V_G - V_T$, to the linear form

$$I_D \simeq \frac{Z}{L}\mu_n C_i (V_G - V_T)V_D \tag{3.42}$$

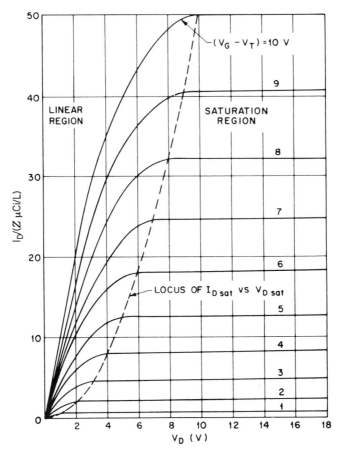

Figure 3.14. Idealized drain (I_D-V_D) characteristics of a MOSFET. The dashed line indicates the locus of the saturation drain voltage ($V_{D\,\text{sat}}$). For $V_D > V_{D\,\text{sat}}$, the drain current remains constant; Z is the channel width in Fig. 3.12 (Sze, 1981).

where V_T is the threshold voltage at which the inversion layer onset occurs:

$$V_T = 2\left(\psi_B + \frac{\sqrt{\epsilon_s q N_A \psi_B}}{C_i} \right) \tag{3.43}$$

The conductance of the MOSFET is zero for V_G lower than V_T. Beyond this value the inversion layer starts to form under the metal gate and the conductance increases linearly with V_G. By plotting I_D vs. V_G (for small and constant V_D), the threshold voltage can be deduced from the linearly extrapolated value at the V_G axis (Lundström and Svensson, 1985). In the linear region [Eq. (3.42)], the channel conductance g_D is given as

$$g_D \equiv \left.\frac{\partial I_D}{\partial V_D}\right|_{V_G = \text{const.}} = \frac{Z}{L}\mu_n C_i(V_G - V_T) \tag{3.44}$$

With sufficient increase of the drain voltage, the charge in the inversion layer at $y = L$ becomes zero, and as a result the number of mobile electrons at the drain falls off drastically [pinch-off point, Fig. 3.13(b)]. The drain voltage and current assume their saturation values, and any V_D increase beyond the pinch-off point results in the I_D being in the saturation region, $I_{D\text{sat}}$, Fig. 3.14. Under the condition $Q_n(L) = 0$, Eq. (3.34) yields the value of $V_{D\text{sat}}$.

$$V_{D\text{sat}} = V_G - 2\psi_B + K^2(1 - \sqrt{1 + 2V_G/K^2}) \tag{3.45}$$

where $K \equiv \sqrt{\epsilon_s q N_A}/C_i$. Now the saturation current can be obtained by substituting Eq. (3.45) into Eq. (3.40):

$$I_{D\text{sat}} \simeq \frac{mZ}{L}\mu_n C_i(V_G - V_T)^2 \tag{3.46}$$

Here m is a function of doping concentration and approaches 1/2 at low dopings (Sze, 1981; Brews, 1981).

In treating real MOSFET characteristics, the first two assumptions of an ideal MIS Schottky diode at the gate and the absence of diffusion currents may be relaxed. The main effect of the fixed oxide charges and the work function difference between the gate metal and the semiconductor is to cause a voltage shift corresponding to the flat-band voltage, V_{FB}, at which the semiconductor surface field and surface charge are zero (Sze, 1981):

$$V_{\text{FB}} = \phi_m - \left(\chi_s + \frac{E_g}{2q} + \psi_B\right) - \frac{Q_F}{C_i} \tag{3.47}$$

for a p-type semiconductor, where ϕ_m is the metal work function; χ_s, the semiconductor electron affinity; E_g, the semiconductor bandgap; ψ_B, the potential difference between the Fermi level E_F and the intrinsic Fermi level E_i; and Q_F, the fixed oxide charge. The V_{FB} shift causes, in turn, a change in the threshold voltage, V_T. In the linear region we obtain [see Eq. (3.43)]

$$V_T = V_{\text{FB}} + 2\left(\psi_B + \frac{\sqrt{\epsilon_s q N_a \psi_B}}{C_i}\right) \tag{3.48}$$

The term ψ_B in Eq. (3.48) represents the potential difference between the flat-band condition and the onset of inversion (onset of a channel), while the third term represents the effects of the doping impurities.

When diffusion effects are taken into account in the expression for the drain current density, we can write

$$J_D(x, y) = q(\mu_n n \mathscr{E} + D_n \nabla n) \tag{3.49}$$

where $n(x, y)$ is the electron number density. The total drain current based on the gradual-channel approximation is (Sze, 1981)

$$I_D = \frac{Z}{L}\left(\frac{\epsilon_s \mu_n}{L_d}\right) \int_0^{v_D} \int_{\psi_B}^{\psi_s} \frac{e^{\beta(\psi - V)}}{F(\beta\psi, V, n_{po}/p_{po})} \, d\psi \, dV \tag{3.50}$$

where β and L_D were defined in Eq. (3.22), and

$$F(\beta\psi, V, n_{po}/p_{po}) \equiv \left[e^{-\beta\psi} + \beta\psi - 1 + \frac{n_{po}}{p_{po}} e^{-\beta V}(e^{\beta\psi} - \beta\psi e^{\beta V} - 1) \right]^{1/2} \tag{3.51}$$

The effect of diffusion current on the gate voltage is a shift related to the semiconductor surface potential ψ_s:

$$V'_G = V_G - V_{FB} = \frac{2\epsilon_s \beta}{C_i L_D} F\left(\beta\psi_s, V, \frac{n_{po}}{p_{po}}\right) + \psi_s \tag{3.52}$$

Equation (3.50) reduces to Eq. (3.40) for gate voltages well above threshold. Although much work has been done in recent years addressing other non-idealities in the basic MOSFET structure (Nicollian and Brews, 1982), they are of no general concern to gas sensor devices and will not be further discussed here.

3.4. SURFACE-FILM DEVICES

The device categories described earlier can also be thought of as relying on thin-film components, such as the surface metal of Schottky diodes or the gate metal and insulator layers in MOSFETs. Thin-film semiconductor substrates have also been used by means of deposition on an insulating sub-strate as early as the 1960s (Weimer, 1962). Crystalline defects and electronic interface traps are more prevalent in these devices than in those with crystalline semiconductors as substrates and are believed to be the main reason for the

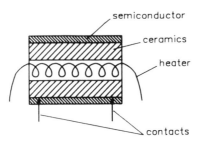

Figure 3.15. A semiconducting gas sensor (Taguchi sensor) heating element (Heiland, 1982).

relative scarcity of gas sensors based on thin-film semiconducting substrates. One area that has witnessed relative success in thin-film gas sensor applications is the use of gas-sensitive semiconductor resistors. These devices almost always use oxide semiconductors in simple structures (Fig. 3.15). The operating mechanism is usually the change of the thin-film conductivity in the presence of certain ambient reducing gases. Specifically, the most important sensing mechanism is likely to be the reaction of the reducing gases with adsorbed oxygen on the surface of, e.g., ZnO or SnO_2 semiconductors. Absorbed oxygen possibly acts as an acceptor and thus induces a decrease in the electron density in the surface region, thus resulting in the lowering of the surface conductivity. Sintered powders or thin films are used because surface and bulk conductivities are always measured in parallel, which for a sensitive measurement requires a large surface-to-bulk ratio (Bergveld and van der Schoot, 1988). Early sensors made of ZnO and SnO_2 oxide semiconductors exhibited resistances strongly dependent on the ambient hydrocarbon vapor concentration (Heiland, 1982). Other metal oxides (e.g., Fe_2O_3, Co_3O_4, and WO_3) have also exhibited similar gas-sensitive properties. The devices themselves consist of thick- or thin-film resistors operating at high temperatures between 150 and 500 °C so as to undergo fast and reversible surface reactions with the gas in question. As shown in Fig. 3.15, the so-called Taguchi sensor (Watson, 1984) includes a heating element for this purpose. It consists of a ceramic cylinder in which there is a heating coil (inside) and a set of electrodes (outside). A thick film of semiconducting oxide is sintered around these electrodes. The sensor can also be made in a flat configuration in which a heater is deposited on one side and the gas-sensitive material on the other side of an alumina substrate (Fukui and Komatsu, 1983). It appears that the Taguchi sensor is more sensitive to carbon monoxide at lower operating temperatures whereas hydrocarbons (e.g., CH_4) can be detected more selectively at high temperatures (Komori et al., 1987).

Despite the high operating temperature and the low selectivity, metal-oxide semiconducting gas sensors are quite popular for qualitative monitoring of combustible gases and as gas leakage detectors, mainly due to their simple fabrication and low cost. To overcome the disadvantage of high operating temperatures of the Taguchi sensor, Wohltjen et al. (1985) have shown that the resistance of thin organic semiconducting films can be changed at room temperature when certain gases (vapors) permeate the bulk of the film. Using the fact that phthalocyanines are electronic semiconductors, these authors made dc measurements of the conductance of modified copper phthalocyanine films that were deposited over an interdigitated electrode array on a quartz substrate by means of the Langmuir–Blodgett technique. In the presence of 0.5–7 ppm NH_3 vapors, the conductivity of the film exhibited bulk behavior: the conductance was shown to be linearly dependent on the number of phthalocyanine monolayers deposited on the sensor. Figure 3.16 shows the electrical circuitry required to fabricate the gas sensor system, including a reference sensor coated with a passivation layer of paraffin wax and used to minimize the strong temperature dependence.

The measured parameter conductance, the reciprocal of resistance, is essentially the surface conductance, since it is unlikely that the conductivity will be homogeneous throughout the (thin or thick) semiconductor oxide film. For this reason only relative conductivities or resistivities are important measurable quantities. The simplicity of the measurement configuration, such as depicted in Fig. 3.16, makes the conductance gas sensor utilization attractive, even though this second-order parameter often renders the signal interpretation in terms of the exact processes occurring during gas–surface reactions quite difficult. Although the original measurements were made, as already discussed, using ZnO (Heiland, 1954; Taguchi, 1970), most commercial gas detectors at present make use of SnO_2 sensing films because they offer high sensitivity at lower operating temperatures (Hagen et al., 1983).

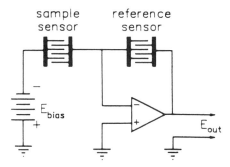

Figure 3.16. Measurement circuitry of the phthalocyanine-based thin-film gas sensor, including temperature compensation (reference resistor). The output voltage, E_{out}, is equal to $-E_{bias}$ multiplied by the conductivity ratio of active and reference sensors (Wohltjen et al., 1985).

Commercial designs generally achieve high surface-to-volume ratios of the active semiconducting elements through the use of thin sputtered films or highly porous aggregates of loosely sintered particles. Introduction of an oxide sensor into an appropriate gas or gas mixture environment perturbs the defect chemistry of semiconducting oxides when the gas mixture contains an oxygen component (Kofstad, 1972). This, in turn, affects the electrical conductivity of the sensor in the presence of an oxygen partial pressure $p(O_2)$. Sensors utilizing TiO_2 have been developed (Tien et al., 1975; Ford Motor Co., 1982) based on this principle. Several physical processes can be involved in registering changes in the composition of a gaseous atmosphere as changes in the conductance of a semiconducting ceramic oxide (Williams, 1987). Figure 3.17 is an illustration of these processes.

Two different classes of operating mechanisms of oxide film semiconducting gas sensors may be distinguished. One class involves changes in the bulk conductance. Generally, they are only sensitive to changes in $p(O_2)$, and the oxide defect chemistry controls the sensor behavior as the film bulk tends toward equilibration with the ambient oxygen pressure. Another class depends on changes in surface conductance and involves small concentration sensing of reactive gases and ambient oxygen in air, through a departure of the value of the conductance from the equilibrium condition. In the case of reactive gases in air, the gas phase is not at equilibrium and the components combust on the surface of the sensor. Therefore, the measured signals depend strongly on diffusion and reaction conditions at or near the interface between the gas and the sensor and within the porous sensor mass (Williams, 1987).

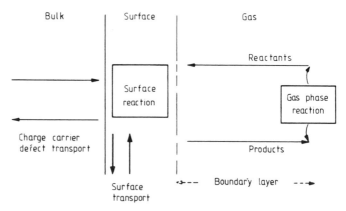

Figure 3.17. Illustration of physical phenomena involved in the transduction of a change in the composition of a gaseous atmosphere into a change in conductance of a ceramic semiconductor oxide. Surface reactions may be between adsorbed species and defects in the solid; between different adsorbed species; between gas molecules and adsorbed species; or any or all of these together. The surface may involve a catalyst (Williams, 1987).

3.4.1. Bulk Conductivity Sensors

The physics and chemistry of bulk conductivity changes in the TiO_2 sensor due to a partial pressure of ambient oxygen represents a pedagogical example of typical operating mechanisms in bulk oxide semiconductor sensors (Logothetis and Hetrick, 1979) and has been studies extensively by Williams (1987). The electrical conductivity is related to the carrier concentration via

$$\sigma = nq\mu_n \qquad (3.53)$$

where μ_n is the electron mobility. In the presence of ambient oxygen of partial pressure $p(O_2)$, reduction processes at the surface of the TiO_2 sensor produce oxygen vacancies, V_O^{2+}, and titanium interstitials, Ti^- (Marucco et al., 1981). The relative importance of these defects in signal generation depends on temperature: oxygen vacancies dominate below ca. 1000 °C. The solid state chemistry is described by the following reactions (Williams, 1987):

$$2Ti^0 + O^0 \rightarrow 2Ti^- + V_O^{2+} + \tfrac{1}{2}O_2\uparrow \qquad (3.54a)$$

$$Ti^- \rightarrow Ti^0 + e^- \qquad (3.54b)$$

$$\overline{O^0 \rightarrow V_O^{2+} + 2e^- + \tfrac{1}{2}O_2\uparrow} \qquad (3.54c)$$

The activation energy for conduction of free electrons consists of the sum of the activation energy, ΔG_2, for reaction (3.54a), which creates donor states Ti^-, i.e., reduced titanium atoms, plus the activation energy, E_D, for reaction (3.54b), which is the amount of energy required for the ionization of a bound electron from donor state Ti^- into the conduction band of the oxide semiconductor. Assuming $E_D \gg kT$ and $n \ll [Ti^-]$, mass-action relations can be written corresponding to the solid state reactions (3.54a) and (3.45b):

$$n = [Ti^-]\exp(-E_D/kT) \qquad (3.55)$$

and

$$[Ti^-]^2[V_O^{2+}]\sqrt{p(O_2)} = \exp(-\Delta G_2/kT) \qquad (3.56)$$

The condition for lattice electrical neutrality is

$$2[V_O^{2+}] = [Ti^-] \qquad (3.57)$$

Combination of Eqs. (3.53)–(3.57) gives the change in conductivity upon removal of oxygen gas from the TiO_2 lattice (Kofstad, 1972)

$$\sigma = \sqrt{2}q\mu_n[p(O_2)]^{-1/6}\exp\{[(\Delta G_2/2) + E_D]/kT\} \qquad (3.58)$$

If $E_D \ll kT$, Eq. (3.58) is obtained with $E_D = 0$.

The case where the lattice is doped with an impurity of different valency has also been studied by Williams (1987), who showed that if trace of Al_2O_3 is doped into TiO_2 it causes a partial oxidation to preserve charge neutrality:

$$V_O^{2+} + 2Ti^- + Al_2O_3 \rightarrow 2Al^- + 2Ti^0 + \tfrac{3}{2}O_2\uparrow \qquad (3.59)$$

where Al^- symbolizes a negatively charged aluminum ion occupying substitutionally a Ti site. The neutrality condition becomes

$$2[V_O^{2+}] = [Al^-] \qquad (3.59a)$$

so that the electrical conductivity of the thin TiO_2 film is lowered:

$$\sigma = \frac{1}{\sqrt{2}}q\mu_n[p(O_2)]^{-1/4}[Al^-]^{-1/2}\exp\{-[(\Delta G_2/2) + E_D]/kT\} \qquad (3.60)$$

If, on the other hand, a trace of higher valence element such as Nb is incorporated, then charge compensation of the Al^- by the positively charged substitutional Nb^+ takes place and the conductivity increases.

Another mechanism contributing to bulk conductivity changes in semiconductor oxides is thermal interband excitation across the bandgap, E_g. For instance, TiO_2 is a wide-bandgap semiconductor ($E_g \simeq 3\,eV$), and

$$n(T)p(T) = K(T)\exp(-E_g/kT) \qquad (3.61)$$

where $p(T)$ denotes the free hole density in the valence band, and $K(T)$ is a preexponential factor, equal to the product of the effective density of states in the conduction and valence bands (Sze, 1981). The electrical conductivity component due to the interband thermal activation process is

$$\sigma = q(\mu_n n + \mu_p p) \qquad (3.62)$$

where μ_p is the hole mobility. For pure materials and at sufficiently high oxygen partial pressure and/or temperature, the conductivity would become independent of $p(O_2)$, as can be seen from Eqs. (3.59) and (3.62).

In doped materials Logothetis and Hetrick (1979) showed that the conductivity exhibits a minimum with increasing oxygen pressure and then an increase with further increases in $p(O_2)$; in this range the nominally n-type TiO_2 behaves like a p-type material. Williams (1987) has shown that in the case of trace Al_2O_3-doped TiO_2 the $p_{min}(O_2)$ corresponding to the conductivity minimum is given by

$$p_{min}(O_2) = \left(\frac{\mu_n}{\mu_p}\right)^{1/2} \left(\frac{2K_2 K_1^2}{K_3 [Al^-]}\right)^2 \tag{3.63}$$

where
$$K_1 = \exp(-E_D/kT)$$
$$K_2 = \exp(-\Delta G_2/kT)$$
$$K_3 = \exp(-E_g/kT)$$

Electrical conductivity expressions such as Eqs. (3.58) and (3.60) are indicative of the need to operate bulk conductivity semiconductor sensors at elevated temperatures for the response time of the sensor to remain within reasonable bounds. An additional problem is the very high sensitivity of the sensor to temperature fluctuations, which may easily overshadow changes in the desired measurand, the $p(O_2)$ in the TiO_2 case just described. Therefore, either very accurate temperature control or the use of reference/compensation devices is needed.

Mainly for this reason much research effort has been directed toward the discovery of materials with higher oxygen sensitivity and lower activation energy than TiO_2. With a lower activation energy the Boltzmann factors in the various expressions for $\sigma(T)$ become less predominant in controlling the sensitivity of the sensor to temperature fluctuations. Such materials include complex oxides (usually perovskites: Williams, 1987), variable-valence transition elements (Mn, Co, Ni: Sekido and Ariga, 1982; Fe: Williams et al., 1984; Cr: Hunter and Brook, 1981), and some simple oxides [CoO: Logothetis et al., 1975; $Co_{1-x}Mg_xO$ ($x < 0.85$): Park and Logothetis, 1977]. Good experimental implementation of such materials as sensors in terms of computer-aided signal acquisition and analysis has greatly contributed toward their utilization despite the sensitive dependence of output signals to temperature fluctuations. At the time of the writing of this book it appears that the substituted alkaline-earth ferrates (Williams et al., 1984) can adequately satisfy the low activation energy condition toward a sensor with high sensitivity to oxygen gas. These materials behave like p-type semiconductors, i.e., the electrical conductivity increases with increasing $p(O_2)$. They exhibit a low *net* activation energy because the creation of a donor state is an exothermic process, the result being that the two energy changes in Eqs. (3.54a, b) partially cancel. The electrical conductivity of the alkaline-earth ferrates can be altered by substituting elements of valence 4 and 5 for iron and elements of valence 3 for the alkaline earth. Figure 3.18 shows typical conductivity and composi-

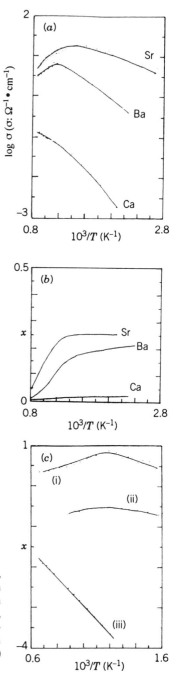

Figure 3.18. Conductivity, σ, and composition increment, x, as functions of absolute temperature, T: (a) comparison of the electrical conductivity and (b) the composition determined by the thermogravimetry of $MFeO_{2.5+x}$, where M = Sr, Ba, or Ca; (c) comparison of the conductivity of successively more highly substituted barium tantalum ferrates, $BaFe_{1-y}Ta_yO_{3-x}$, for (i) $y = 0$, (ii) $y = 0.25$, and (iii) $y = 0.5$ (Williams, 1987).

tion dependences on temperature for several among these sensor materials. In the case of barium ferrate, the composition is a function of temperature because oxygen is lost from oxidized material upon temperature increase. Thus, the oxidized p-type material has higher conductivity than reduced n-type material. As a result a fortuitous cancellation of the effects on the conductivity occurs, which leads to the maximum observed for $BaFeO_{3-x}$ in Fig. 3.18(a). Similar processes are responsible for the smoother maximum of the $SrFeO_{3-x}$ curve. This class of materials exhibits optimum sensitivity to $p(O_2)$ owing to their near-flat temperature dependence, which is partly due to the aforementioned mechanisms and partly due to further smoothing out as a consequence of partial reduction caused by tantalum substitution (Figure 3.19).

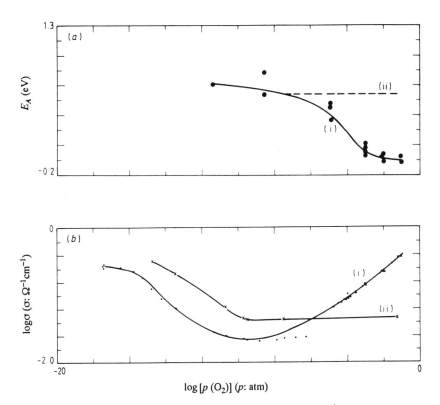

Figure 3.19. (a) Dependence of conductivity activation energy over the temperature range 920–1120 K; and (b) the isothermal ($T = 1100$ K) conductivity dependence on oxygen partial pressure for (i) $BaFe_{0.5}Ta_{0.5}O_3$ and (ii) $BaFe_{0.66}Ta_{0.33}O_{3-x}$ (Williams, 1987).

3.4.2. Surface Conductivity Sensors

This category of sensors includes thin-film devices that exhibit an electrical conductive response as the result of a surface charge-exchange process with an ambient gaseous component adsorbed on the surface of the active layer. This general definition seems to best represent the response of SnO_2 and ZnO to NO_2 and other strongly oxidizing species (Williams, 1987). A related model assuming the displacement of chemisorbed oxygen by the gas has also been used extensively, with one or the other model being successful in explaining the sensor response in the majority of cases (Stoneham, 1987). In what follows we shall invoke the model used by Windischmann and Mark (1979) to describe the mechanism of thin-film devices with an active layer of semi-conducting SnO_2 (100–1000 Å) deposited on a suitable insulating substrate. Figure 3.20 shows the energy-band structure of the device. The criterion of the layer being defined as a "thick film" or a "thin film" is the relative value of its thickness and the extent of the depletion layer, W_{sc}. This layer is depleted

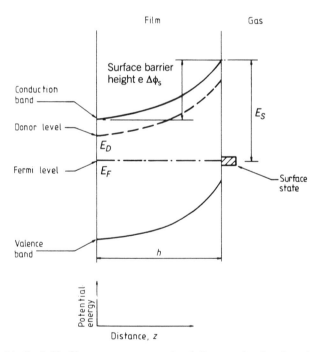

Figure 3.20. Idealized thin-film gas sensor energy-band diagram, showing the principal energy states and defining the notation used in the text. The direction of current flow is parallel to the film–gas interface (Windischmann and Mark, 1979).

of conduction electrons owing to efficient trapping is surface states (Madelung, 1978). The depletion layer width, W_{sc}, and the surface potential (Schottky barrier) height, $\Delta\phi_s$, both depend on the amount of charge stored in the surface states. If $n(z)$ denotes the conduction electron density along the direction z normal to the film–gas interface (Fig. 3.20), then the conductivity of the layer may be calculated using several simplifying assumptions by introducing the dependence of $n(z)$ on the local electrostatic potential, $V(z)$. Poisson's equation then gives the relationship between $V(z)$ and the local charge density, $\rho(z)$. In the case of surface-trap-limited conductance of a fully depleted thin layer, one obtains

$$\sigma = q\mu_n nh \tag{3.64}$$

along with the mass-balance condition (Williams, 1987) valid for a single-sided channel:

$$N_D h = N_s \theta \tag{3.65}$$

where $N_D (N_s)$ is the ionized donor (surface) state density (number of states per unit volume) and θ is the surface state occupancy. For a generalized reaction scheme at the surface

$$S + e^- + \frac{m}{2} O_2 \underset{K_{-1}}{\overset{K_1}{\rightleftharpoons}} O_m^- \tag{3.66a}$$

$$O_m^- + mR \xrightarrow{K_2} mRO + e^- \tag{3.66b}$$

where S denotes a surface adsorption site for oxygen; e^-, a conduction electron; and R, a combustible gas that reacts with adsorbed oxygen species and establishes a steady state occupancy of the surface states less than that in air. If θ is the coverage of the surface oxygen, O_s^-, then at steady state $d\theta/dt = 0$, so that (Williams, 1987)

$$\theta_{ss} = K_1 n[p(O_2)]^{m/2}/(K_{-1} + K_2[p(R)]^m) \tag{3.67}$$

Desorption involves activation of an electron from a surface state (activation energy E_s); adsorption involves ionization of a donor (activation energy E_D). In the scheme of Eqs. (3.66a, b) the rate constants

$$K_1(T) = K_1^0 \exp(-E_D/kT) \tag{3.68}$$

$$K_{-1}(T) = K_{-1}^0 \exp(-E_s/kT) \tag{3.69}$$

and

$$K_2(T) = K_2^0 \exp(-E_s/kT) \tag{3.70}$$

Combination of Eqs. (3.65) and (3.67) gives n in terms of $p(O_2)$, K_1, K_2, and $p(R)$, while Eqs. (3.64), (3.68), and (3.70) finally give the expression for the sensor conductance corresponding to the surface reaction scheme (3.66a, b):

$$\sigma = (q\mu_n N_D h^2/N_s K_1^0)\{\exp[-(E_s - E_D)/kT]\}$$
$$\cdot [p(O_2)]^{-m/2}(K_{-1}^0 + K_2^0[p(R)]^m) \tag{3.71}$$

Equation (3.71) shows that the conductance activation energy is $(E_s - E_D)$ and that the reaction orders with respect to O_2 and R are $-m/2$ and m, respectively. Baidyaroy and Mark (1972) have treated the case where the surface states have a distribution of energies. Strassler and Reis (1983) have studied more complex reaction schemes than Eqs. (3.66), involving a multiplicity of differently charged oxygen species.

In the case of barrier-limited conductance, the charge transport across a surface (Schottky) barrier is considered (Fig. 3.20):

$$\sigma = \sigma_0 \exp(-q\Delta\phi_s/kT) \tag{3.72}$$

and the first objective is to calculate $\Delta\phi_s$ as a function of gas partial pressures, with the undepleted bulk of the semiconductor supplying the source of electrons for the surface states:

$$n = N_D \tag{3.73}$$

Unlike earlier considerations that led to Eq. (3.68), adsorption here requires donor ionization and electron activation across the surface barrier $E_s = E_D + q\Delta\phi_s$ (Fig. 3.20):

$$K_1 = K_1^0 \exp[-(E_D + q\Delta\phi_s)/kT] \tag{3.74}$$

Equations (3.69) and (3.70) remain valid, and Eq. (3.67) is replaced by

$$\theta_{ss} = \frac{K_1^0 N_D \exp[(E_s - E_D)/kT]\exp(-e\Delta\phi_s/kT)[p(O_2)]^{m/2}}{(K_{-1}^0 + K_2^0[p(R)]^m)} \tag{3.75}$$

The depletion layer width is now given by the following expression when only one type of surface states exists with charge q per unit area and fractional occupancy θ_{ss} (Williams, 1987):

$$W_{sc} = \frac{Q_s}{qN_D} = \frac{N_s q \theta_{ss}}{N_D}$$ (3.76)

The Schottky barrier height is [Eq. (3.9)]

$$\Delta\phi_s = \frac{Q_2^2}{2q\epsilon_s N_D} = \frac{qN_s^2 \theta_{ss}^2}{2\epsilon_s N_D}$$ (3.77)

Therefore, upon defining $P \equiv q\,\Delta\phi_s/kT$ we obtain from Eqs. (3.75) and (3.77):

$$\ln P + 2P = \frac{2(E_s - E_d)}{kT} + \ln\left(\frac{q^2 N_s^2 N_D}{2\epsilon_s kT}\right) + m\ln[p(O_2)]$$
$$+ 2\ln\left(\frac{K_1^0}{K_{-1}^0 + K_2^0[p(R)]^m}\right)$$ (3.78)

Under normal operating conditions $\Delta\phi_s \simeq 0.2\text{--}1.0\,\text{eV}$ and $T \simeq 800\,\text{K}$, it can be shown that $\ln P \ll 2P$, or empirically (McAleer et al., 1986)

$$\ln P + 2P = 2.18P + 0.56$$ (3.79)

Substitution of Eq. (3.79) into (3.78) yields

$$P = \text{const.} + (m/2.18)\ln[p(O_2)] - 0.92\ln(K_1^0 + K_2^0[p(R)]^m)$$ (3.80)

which shows a semilogarithmic relationship between activation energy, $e\,\Delta\phi_s$, and oxygen partial pressure, as well as lowering of the activation energy upon introduction of a reactive gas, R (McAleer et al., 1986). Now, from Eqs. (3.72) and (3.80) one obtains for the sensor conductivity:

$$\sigma \propto \frac{(K_1^0 + K_2^0[p(R)]^m)^{0.92}}{[p(O_2)]^{m/2.18}}$$ (3.81)

Equations (3.71) and (3.72) have a close resemblance and are distinguished only by their different predictions regarding the conductance activation energy. Furthermore, if the approximation $\ln p \ll 2p$ is made, valid for large $\Delta\phi_s$, then Eqs. (3.75) and (3.78) yield the estimate

$$\theta_{ss} = N_D \left(\frac{2\epsilon_s kT}{q^2 N_s^2 N_D}\right)^{1/2} \sim 10^{-3}$$ (3.82)

while Eq. (3.76) gives

$$W_{sc} = \left(\frac{2\epsilon_s kT}{q_2 N_D}\right)^{1/2} \sim 10 \, \text{nm} \tag{3.83}$$

3.5. CLASSIFICATION AND PERFORMANCE OF SEMICONDUCTOR GAS SENSOR DEVICES

3.5.1. Classification Scheme for Semiconductor Gas Sensors

Saaman and Bergveld (1985) have given a classification of XIS systems, in which S stands for semiconductor, X is either a conductor (electronic or ionic) or a gas, and I is an electrical insulator. Figure 3.21 is a visual review of the portion of their classification scheme that is relevant to gas sensor devices. This scheme is based on analogies (a) between related fields of semiconductor research and (b) between gas sensors and their chemically insensitive microelectronic equivalents. Three cases can be distinguished, depending on the thickness of the insulating layer I: (1) The insulating layer is of high enough quality and sufficiently thick so as to be an insulator in the conventional sense (denoted as "I" in MIS); (2) the insulator thickness is less than 5 nm, so that quantum mechanical tunneling phenomena across the insulator dominate the electronic conduction processes (such a layer is denoted as "i" in MiS); (3) the insulating layer is absent. It is these considerations which led to the classification scheme shown in Fig. 3.21. Important representative devices draw strong analogies on the fundamental building-block microelectronic device structures (Schottky barriers and MOS structures) presented earlier in Sections 3.2 and 3.3.

3.5.1.1. MIS–Chemical Gas Sensors

In these thick-insulator-layer devices the conventional gate metal (normally Al) has been replaced by metals known to exhibit chemical reactivity, such as the group VIII elements Ni, Pd, and Pt. MIS devices have thus been fabricated exhibiting sensitivity to hydrogen (Lundström et al., 1975a, 1977), carbon monoxide (Krey et al., 1983), hydrogen disulfide (Shivaraman, 1976), and ammonia (Lundström et al., 1975b), among other gases. The operation of all these sensors originates on the formation of a dipole layer at the metal–insulator interface. The resulting voltage drop in the presence of a chemically active ambient gas is measured as a flat-band voltage shift in both MIS capacitors and MISFETs. This voltage drop may also be described in terms of the metal work function dependence on adsorbed/absorbed chemically active gas concentrations. An interesting variant of these devices with strong

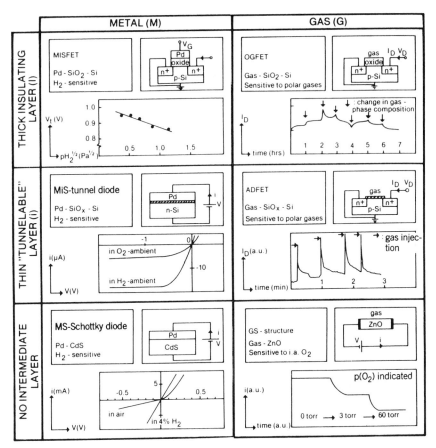

Figure 3.21. Classification of chemically sensitive semiconductor gas sensors. For each category a representative device structure is depicted and its electrical output (a characteristic current or voltage) is shown schematically as a function of the input gas partial pressure. Data are taken from the literature; see text for details (Saaman and Bergveld, 1985).

gas-sensing potential is the Hg-gate MIS capacitor described by Corker and Svensson (1978) as sensitive sensors upon the addition of solid Na to the Hg. These authors observed large flat-band voltage shifts of this device attributed to the formation of sodium amalgam, the work function of which is quite different from that of pure Hg.

3.5.1.2. MiS–Chemical Gas Sensors

Representative examples of these sensors characterized by thin insulator layers are the Pd–SiO_x–Si hydrogen sensor (Zemel et al., 1981) and the

Al–SiO$_x$–Si moisture sensor (Duszak et al., 1981). The device by Zemel et al. (1981) operates as a MiS tunnel diode, whereas Duszak and co-workers' device operates on the transistor principle, employing two interacting MiS tunnel diodes. Although the fundamental mechanisms underlying these sensors are not yet fully elucidated, it is clear that the role of the thin insulator is different in each case: in the moisture sensor the SiO$_x$ layer is likely to be involved in the chemical aspect of sensing, whereas its main role in the hydrogen sensor is to prevent the undesired formation of intermediate species such as Pd$_2$Si while preserving device sensitivity to hydrogen. D'Amico et al. (1983a, b) have further demonstrated the sensitivity to hydrogen gas of a Pd-coated amorphous hydrogenated silicon (Pd/a-Si:H) MiS Schottky barrier structure.

3.5.1.3. *Metal–Semiconductor Chemical Gas Sensors*

Metal–semiconductor (Schottky diode) junctions have been used toward gas sensor fabrication with Pd as the most frequent catalytic active metal. In order for a successful sensor to be fabricated, the semiconducting substrate must satisfy two essential requirements (Saaman and Bergveld, 1985): no intermediate chemical species must be formed through interaction with the metal, and the Fermi level of the semiconductor must not be pinned due to the existence of large interface state distributions. Hydrogen sensors have thus been reported using several Schottky diode systems. These include the Pd–ZnO (Ito, 1979), Pd–CdS (Steele and MacIver, 1976) and Pd–TiO$_2$ (Yamamoto et al., 1980) sensors.

3.5.2. MIS Capacitor Chemical Gas Sensors

3.5.2.1. *Devices, Mechanisms, and Applications*

This is perhaps the simplest semiconductor-based gas sensor device to fabricate. Lundström et al. (1977) reported such a device using *p*-type silicon substrate, which was first oxidized in dry oxygen at 1200 °C to grow a thickness of 1000 Å oxide. Subsequently, the oxide was etched away from the back of the wafer, followed by aluminum evaporation, which may be sintered. Palladium metal was evaporated on the front side of the wafer through a mask and produced Pd dots of approximately 1 mm^2 area. The Pd layer is typically 50–100 nm thick and is thermally evaporated on the SiO$_2$. Unfortunately, thin Pd films have very weak adhesion to silicon dioxide. Palladium has been known to peel off during fabrication (Shivaraman and Svensson, 1976) and to form blisters after exposure to hydrogen gas (Armgarth and Nylander, 1982). A solution to these problems has been given through the use of heat treatment in air at 200 °C following fabrication (Shivaraman and Svensson, 1976), and the deposition of thicker Pd films (> 400 nm) (Armgarth and Nylander, 1982). A complete sensor is made when a heater is incorporated with the capacitor, along with a temperature sensor to keep the device at constant temperature. During the measurement a capacitance meter is used to track changes in the capacitance. Two measurement modes are possible: one either uses the capacitance change as a measure of gas concentration (Steele et al., 1976), or one keeps the capacitance constant by varying the bias voltage, which in turn is used as a measure of gas concentration (Lundström et al., 1977).

Lundström et al. (1975a) have introduced the hydrogen-sensitive MIS capacitor and have explored its sensitivity to hydrogen gas as shown in Fig. 3.22. Hydrogen molecules in air are dissociated on the Pd catalytic metal

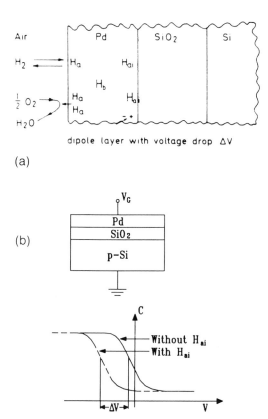

Figure 3.22. Schematic diagram of hydrogen-sensitive Pd–MOS capacitor: (a) principle (see text); (b) a Pd–MOS capacitor and its $C-V$ curve (Lundström, 1981).

surface, and the atoms are subsequently adsorbed on the metal surface. Some of these atoms diffuse through the metal film and reach the metal–insulator interface. The number of adsorbed hydrogen atoms on the front (Pd–gas) interface is a function of the hydrogen pressure in the chamber, as well as of the presence of the other gases in the ambient. In an inert atmosphere such as argon or nitrogen, the only reactions taking place on the Pd surface are the dissociation and association of hydrogen (Lundström, 1981):

$$H_2 \underset{d_1}{\overset{c_1}{\rightleftarrows}} 2H_a \tag{3.84}$$

where H_a stands for adsorbed hydrogen on the surface, and c_1 and d_1 are rate constants for the reaction. The adsorbed hydrogen, H_a, communicates with the bulk of the metal film and with the Pd–SiO$_2$ interface (Fig. 3.23):

$$H_a \underset{d_e}{\overset{c_e}{\rightleftarrows}} H_b \underset{c_i}{\overset{d_i}{\rightleftarrows}} H_{ai} \tag{3.85}$$

For the definitions of the various symbols, the reader is referred to Fig. 3.23. There appears to be no time delay between gas–metal surface and metal–oxide

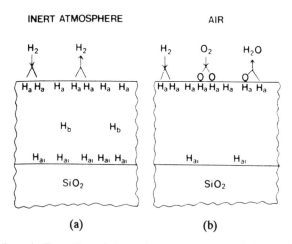

Figure 3.23. Schematic illustration of chemical reactions on a catalytic metal surface in the presence of hydrogen. (a) In an inert atmosphere only dissociation and association of hydrogen takes place: [H_a] is the concentration of adsorbed hydrogen atoms on the surface; [H_b] is that in the metal; and [H_{ai}] is that at the Pd–SiO$_2$ interface. (b) In oxygen, production of H$_2$O takes place, thus decreasing the number of available hydrogen atoms and hence [H_{ai}] (Lundström, 1981).

interface events on the time scale of all reported MOS capacitor or MOS transistor experiments, which occur at ca. $100\,°C$ with resolution $\sim 1\,s$. Assuming that the number of hydrogen adsorption sites on the metal surface is N_e and on the metal–insulator interface is N_i, Lundström (1981) wrote the kinetic equations for hydrogen transport in the Pd bulk and derived the equilibrium conditions:

$$\frac{\theta_i}{1 - \theta_i} = \left(\frac{c_e d_i}{d_e c_i}\right) \frac{\theta_e}{1 - \theta_e} = \left[\frac{k_i c_1}{d_1} p(\mathrm{H_2})\right]^{1/2} \tag{3.86}$$

where θ_i (θ_e) is the coverage of hydrogen at the interface (surface): $\theta_j \equiv n_j/N_j$; $j = i, e$. The constant $k_i = (c_e d_i/d_e c_i)^2$ depends mainly on the difference in adsorption energies at each interface. It appears to have a value close to 1 for the adsorption sites for hydrogen on Pd. Equation (3.86) leads to a Langmuir isotherm, with a voltage drop across the MOS capacitor given by

$$\Delta V = \Delta V_{\max} \theta_i \tag{3.87}$$

where $\Delta V_{\max}$ is the maximum shift of the MIS characteristics for fully saturated adsorption sites, i.e., for $\theta_i = 1$. Figure 3.24 shows typical experimental capacitor sensor results for hydrogen in argon, with the parameter c_1/d_1 in Eq. (3.86) having the temperature dependence

$$\frac{c_1}{d_1} \propto \exp(q E_a/kT) \tag{3.88}$$

where E_a is the adsorption energy per hydrogen molecule or $2\mathrm{H}_a$. In Fig. 3.24(b) the experimental results are plotted upon combining Eqs. (3.86) and (3.87) in the form

$$\frac{1}{\Delta V} - \frac{1}{\Delta V_{\max}} = \frac{1}{\Delta V_{\max}} \left(\frac{d_1}{k_i c_1 p(\mathrm{H_2})}\right)^{1/2} \tag{3.89}$$

In air, the following reaction scheme has been proposed for hydrogen sensing (Lundström et al., 1975a; Plihal, 1977):

$$\mathrm{H_2} \xrightarrow{c_1} 2\mathrm{H}_a \tag{3.90}$$

$$\mathrm{O_2} + 2\mathrm{H}_a \xrightarrow{c_2} 2(\mathrm{OH}_a) \tag{3.91}$$

$$\mathrm{OH}_a + \mathrm{H}_a \xrightarrow{c_3} \mathrm{H_2O} \tag{3.92}$$

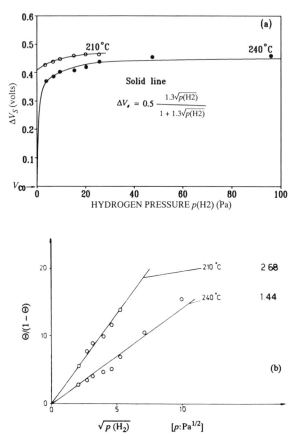

Figure 3.24. (a) Typical experimental results for hydrogen in argon at two temperatures, obtained by use of a MOS capacitor sensor. The response becomes saturated at relatively low $p(H_2)$. The solid line is calculated according to Eqs. (3.86) and (3.87). (b) The experimental results plotted according to Eq. (3.89). In this case $\Delta V_{max} = 0.5$ V was used as determined from (a) (Lundström, 1981).

The back-reactions are assumed to be inefficient and are therefore neglected. The steady state coverage of hydrogen at the surface is given by

$$\frac{\theta_i}{1 - \theta_1} = \left[\frac{k_i c_1}{2c_2} p(H_2) \right]^{1/2} \tag{3.93}$$

in agreement with experimental isotherms (Lundström et al., 1975a; Plihal, 1977). In general, it appears that the voltage shift of the MOS capacitor sensor in air follows isotherms of the form

$$\frac{\theta_i}{1 - \theta_i} \propto [p(\mathrm{H}_2)]^{1/2} \qquad \text{(hydrogen dependence)}$$

and

$$\frac{\theta_i}{1 - \theta_i} \propto \frac{1}{[p(\mathrm{O}_2)]^{1/2}} \qquad \text{(oxygen dependence)}$$

The sensitivity of the sensor is much lower in air than in an inert atmosphere owing to chemical reactions on the metal surface. Lundström (1981) has shown that the presence of hydrogen in the Pd matrix and along the metal–insulator interface sets up a dipole layer that induces a voltage drop ΔV given by (see Fig. 3.23)

$$\Delta V = Np/\epsilon_0 \qquad (3.94)$$

where N is the density of the absorbed hydrogen atoms, and p is the dipole moment of the atom. It has been assumed that the surroundings of each dipole consists of vacuum, represented by the dielectric constant ϵ_0. In the presence of the interfacial dipole layer the voltage drop ΔV shifts the flat-band voltage [Eq. (3.47)]

$$V_{\mathrm{FB}} = (V_{\mathrm{FB}})_0 - \Delta V \qquad (3.95)$$

as well as the height of the energy barrier between the metal and the insulator $\phi_{mB} = \phi_m - \phi_B$ (Fig. 3.25):

$$\phi_{mB} = (\phi_{mB})_0 - q\,\Delta V \qquad (3.96)$$

Experimentally it is found that $\Delta V_{\max} \simeq 0.5\,\mathrm{V}$. If this corresponds to an interface with one hydrogen atom per palladium atom, with $N_{\mathrm{Pd}} \simeq 2 \times 10^{19}\,\mathrm{m}^{-2}$, Eq. (3.94) gives $p = 2.2 \times 10^{-31}\,\mathrm{A \cdot s \cdot m}$, which implies a separation of ca. $0.014\,\text{Å}$ between the positive and the negative charges of the dipole layer (Lundström, 1981). This situation arises in numerous experimental cases. At ambient temperatures above 50–60 °C the hydrogen and oxygen concentration dependence of ΔV is found to obey the relation (Lundström and Söderberg, 1981/82)

$$\frac{1}{\Delta V} - \frac{1}{\Delta V_{\max}} = \frac{1}{\Delta V_{\max}} \left[\frac{\alpha[p(\mathrm{O}_2)] + \beta}{p(\mathrm{H}_2)} \right]^{1/2} \qquad (3.97)$$

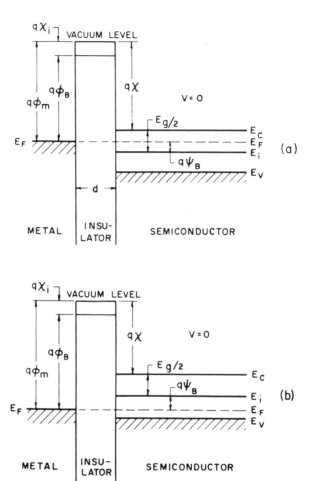

Figure 3.25. Energy-band diagrams of ideal MIS structures at applied bias $V = 0$: (a) n-type semiconductor; (b) p-type semiconductor. Here ϕ_m is the metal work function; χ (χ_i), the semiconductor (insulator) electron affinity; E_g, the bandgap; ϕ_B, the potential barrier between the metal and the insulator; and ψ_B, the potential difference between the Fermi level, E_F, and the intrinsic Fermi level, E_i (Sze, 1981).

in certain regions of hydrogen and oxygen pressures. In Eq. (3.97), α and β are empirical constants determined through curve fitting. Figure 3.26 shows "isopotential" curves constructed from plots of the MIS capacitor sensor potential vs. $p(H_2)$ and $p(O_2)$ such that the connection between hydrogen and oxygen pressure yields a constant measured voltage. The linear behavior

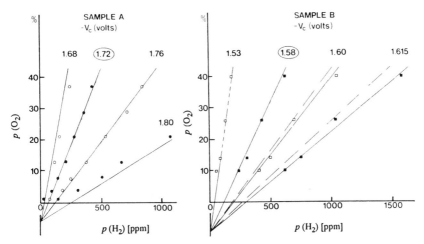

Figure 3.26. Isopotential curves, i.e., curves connecting hydrogen and oxygen pressures yielding the same V_c at $T = 152\,°C$. The straight lines are calculated from Eq. (3.98), where α and β were calculated from the line belonging to the encircled potential. The solid lines for sample B are for $V_{cd} = -1.30\,V$, and the dashed lines for $V_{cd} = -1.36\,V$ (Lundström and Söderberg, 1981/82).

of the $p(O_2)$ vs. $p(H_2)$ plots is expected from the rearrangement of Eq. (3.97):

$$p(H_2) = \left[\frac{\Delta V/\Delta V_{max}}{1 - (\Delta V/\Delta V_{max})} \right]^2 [\alpha p(O_2) + \beta] \tag{3.98}$$

Owing to high differential sensitivity of the sensor to hydrogen at low concentrations, the flat-band voltage condition for the Pd–MOS structure is difficult to achieve and $(V_{FB})_0$ in Eq. (3.95) is therefore normally obtained using extrapolation of the data at very low $p(H_2)$. Such extrapolations for two samples A and B (Fig. 3.26) have given $(V_{FB})_0 \equiv V_{cd} = -1.1\,V$ (sample A) and $V_{cd} = -1.30\,V$ (or $-1.36\,V$) (sample B).

The utility of isopotential plots such as those of Fig. 3.26 is related to recent developments in the direction of multi-gas-sensing pattern recognition instrumentation (see Section 3.5.3.3, below). Lundström and Söderberg (1981/82) have utilized a Kelvin probe consisting of an aged Ni foil of metallurgical grade, as shown in Fig. 3.27, in an effort to separate out the effects of chemical reactions at the Pd surface from the electrostatic interactions occurring at the Pd–insulator interface. The Kelvin probe, a vibrating capacitor, has been extensively used for adsorption studies at surfaces (Engelhardt et al., 1977). Their results with the hydrogen–oxygen system can be summarized along the following lines. Hydrogen is dissociated on the Pd surface and adsorbed

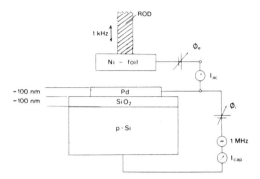

Figure 3.27. The Pd–SiO$_2$–Si test structure used by Lundström and Söderberg (1981/82), and the Kelvin probe attached above it. The measurement methodology for the surface potential ϕ_e and the interface potential ϕ_i is also shown.

on the surface, at the Pd–SiO$_2$ interface, and to some extent in the bulk. The relative importance of these processes is intimately linked with the degree of porosity of the Pd film. There appear to exist some long-term effects during signal transients related to the "spillover" of hydrogen into the oxide, while a degree of quasi-equilibrium is maintained between the surface and the interface during the transient. The quality of the oxide seems to affect the long-term effects, specifically the Na$^+$ ion content in SiO$_2$.

Spillover is well known in heterogeneous catalysis and can occur not only between the Pd-absorbed/SiO$_2$ system but quite often between hydrogen and a metal catalyst, as a process where the catalyst dissociates the molecule into atoms that subsequently "spill over" onto the surface of the substrate if the catalyst is porous or grainy (Bianchi et al., 1975; Sancier, 1971; Sermon and Bond, 1973). Such spillover of oxygen has not been widely reported (Morrison, 1987). An important effect due to spillover is the acceleration of the attainment of steady state conditions in the semiconductor gas sensor. The existing evidence in an inert argon atmosphere regarding the binding of hydrogen in the Pd metal of the MOS capacitor sensor converges toward distributions of deep and shallow binding energy levels (Lundström, 1981; Fortunato et al., 1989). Deep hydrogen adsorption sites with large adsorption energy appear to be filled all the time even under ambient conditions, owing to the presence of trace amounts of hydrogen in the atmosphere. Shallow adsorption sites with small binding energies can be occupied or empty depending on $p(H_2)$ in argon, and adsorbed hydrogen in the Pd bulk can be observed. Fortunato et al. (1989) have studied some of these processes in Pd–SiO$_2$–Si capacitors using optical and electrical transport measurements. They concluded that at low $p(H_2)$, away from the $\alpha \rightarrow \beta$ transition (Lewis,

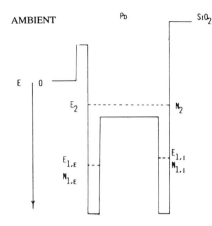

Figure 3.28. Energy level scheme for hydrogen atoms in the Pd–SiO$_2$ system: external surface sites, $N_{1,e}$, with relative energy, $E_{1,e}$; interfacial sites, $N_{1,i}$, with relative energy, $E_{1,i}$; and bulk sites, N_2, with relative energy, E_2 (Fortunato et al., 1989).

1967), both optical and electrical signals are consistent with an energy distribution model involving two different sites for the adsorbed/absorbed hydrogen atoms (Fig. 3.28).

In oxygen more complex phenomena occur, with the overall effect that oxygen molecules react with adsorbed hydrogen atoms on the Pd surface to generate water. The transient response of the Pd–SiO$_2$–Si device in oxygen–hydrogen ambients suggests that the dissociation of oxygen molecules on the Pd surface blocks hydrogen adsorption sites (Lundström and Söderberg, 1981/82). Figure 3.29 schematically shows an important property of the steady state response of the Pd-coated MIS sensor. The Pd layer acts as an effective filter for hydrogen atoms. The coverage of hydrogen at the interface is determined by the effective coverage at the surface, i.e., by the $[H_a]$ and $[OH]$ (Lundström et al., 1989), which means that blocking events at the surface change the adsorbed surface density of hydrogen but not the reaction pathways on the surface. Therefore, the speed of response will change but *not* the steady state value. Thus, in principle, one active site is capable of filling the interface and giving steady state signals the same as those obtained from a "clean" surface, albeit with a much longer transient response. Lundström and Söderberg (1981/82) have shown that under normal operation (100–150 °C) in air or oxygen the bulk properties of the MOS capacitor sensor, such as the resistivity of the metal, are influenced negligibly by the dissolution of hydrogen in Pd. On the other hand, for hydrogen in argon detectable changes in the bulk resistance of the Pd gate have been observed.

The MOS capacitive structure has been utilized as a sensor of several gases besides hydrogen. Table 3.1 shows some examples of reported applications with catalytic metal gates, as collected by Lundström et al. (1986a). In fact, the Pd-gate capacitor is sensitive to all molecules that can donate

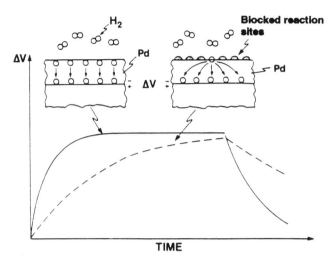

Figure 3.29. Schematic illustration of an interesting property of catalytic metal gates. The steady state response will be the same, even if some of the reaction sites are blocked. The response is, however, much slower for a large number of blocked surface sites (Lundström et al., 1989).

hydrogen atoms to the Pd layer, following surface chemical reaction. Sensitivity has thus been demonstrated upon exposure to hydrogen sulfide, ethanol and other alcohols, unsaturated hydrocarbons, and ammonia. A capacitive response to this last-named molecule was discovered by Winquist et al. (1983). The sensor response manifested itself as a shift of the $C-V$ curve of the device, schematically shown in Fig. 3.30. The exceptional feature of the NH_3 sensor is that the contact metal does not have to be catalytic (Spetz et al., 1987). Aluminum contacts in the 4–63.5 nm thickness range have shown sensitivity to a few ppm of NH_3 in 20% O_2 in N_2 (synthetic air). In comparison, a device with a thick, unmodified Pd gate has a response of ca. 20 mV to 500 ppm NH_3 in air, at a sensor temperature of 150 °C (Lundström et al., 1986a).

It has been found that when a catalytic or noncatalytic metal layer is used as the active element with MIS capacitor sensors of NH_3, the thickness dependence of the device sensitivity is such that it exhibits a broad maximum; for Pt layers this maximum occurs around 10–30 nm thickness (Spetz et al., 1987). Platinum layers thicker than about 90 nm are virtually insensitive to ammonia. It is necessary for the metal to be thick enough to have sufficiently low electrical resistivity and, at the same time, thin enough to be porous. When a discontinuous metal gate (e.g., Pt) is used, a strong enhancement of sensor sensitivity occurs, with shifts of the $C-V$ curve of the order of 500 mV for 50 ppm NH_3 in air (Fig. 3.31) (Winquist et al., 1983; Spetz et al., 1987,

1988; Ross et al., 1987). These observations recently led to a tentative model for ammonia sensitivity, which is not attributed to hydrogen atoms/dipoles produced from catalytic reactions on the metal surface (Lundström et al., 1989), unlike earlier hypotheses (Winquist et al., 1983, 1985a; Ross et al., 1987). According to the recent model, the temperature and oxygen dependence of the ammonia response are consistent with the surface reactions determining the *size* of the voltage shift in the $C-V$ curve. The sensitivity can be attributed to a surface potential change of the metal, capacitively coupled through the holes (pores) in the discontinuous metal. In the schematic notation of Fig. 3.32, the detected voltage shift is given by (Lundström et al., 1989)

$$\Delta V = \Delta V'_s \left[\frac{C_s C_E}{C_s C_E + C_M(C_E + C_s)} \right] \tag{3.99}$$

where C_E and C_M are, to first approximation, geometrical capacitances. The stray capacitance C_s is a function of the thickness and structure of the metal film. It is probably much larger than C_E for very thin metal films, yielding

$$\Delta V \simeq \Delta V'_s \left(\frac{C_E}{C_M + C_E} \right) \tag{3.100}$$

where $\Delta V'_s$ is the average "far field" surface potential change from the metal islands, which may decrease from that of the homogeneous metal, ΔV_s, when the coverage of the metal on the surface decreases. When the metal film thickness increases, C_s becomes smaller than C_E and C_M. Then Eq. (3.99) gives

$$\Delta V \xrightarrow[C_s \to 0]{} \Delta V'_s \left(\frac{C_s}{C_M} \right) \to 0 \tag{3.101}$$

This trend shows the same qualitative dependence on the metal film thickness as the experimental results of Fig. 3.31. On the contrary, the voltage shift of thin Pt–MOS capacitor sensors caused by H_2 is much smaller than that caused by NH_3 and does not show thickness dependence of Fig. 3.31.

The MOS capacitor sensor sensitivity to other gases containing ammonium (NH_4^+) species has been documented in flow-through systems of biological solutions, supplied with a separate enzyme column (Lundström et al., 1986a). If the dilution of the sample is suitable, all the substrate in the sample will be converted to ammonia. This means that for the enzyme column–flow-through cell combination, the pulse response is a direct measure of the amount of (convertible) substrate in the sample (Fig. 3.33). This has been experimentally verified in a number of cases and is summarized in Table 3.2.

Table 3.1. Some Examples of Reported Responses of Gas Sensors with Catalytic Metal Gates

Sensor Type	Operating Temp. (°C)	Reference Gas	Test Gas	Response (mV)	Notes	Ref.
Pd–MOSFET	150	Air	0.005 ppm in air	1	Estimated detection limits	Armgarth et al. (1985)
	150	Air	0.0003 ppb H_2 in argon	1		
Pd–MOSFET	150	Air	50 ppm H_2S in air	30		Poteat et al. (1983)
Pd–MOSFET	150	Air	2.5 ppm H_2S in air	300–400	Slow recovery at 150 °C; fast at 200 °C	Poteat and Lalević (1981)
Pd–MOSFET	150	Air	500 ppm H_2 in air	380	Test of double metal gates	Ruths et al. (1981)
Pt/Pd–	150	Air	500 ppm H_2 in air	110		
Pd/Pt–	150	Air	500 ppm H_2 in air	380		

Pt–	150	Air	500 ppm H_2 in air	110	Ackelid et al. (1985)
Pd–Al$_2$O$_3$–SiO$_2$–Si CAP	75	Air	50 ppm H_2 in air	220	Ackelid et al. (1986)
Pd–Si$_3$N$_4$–InP CAP	Room	Air	760 torr H_2	780	
Pd–InP	Room	Air	100 ppm H_2 in N_2	140	
Schottky diode	Room	Air	760 torr H_2	190	
Pd–MOS CAP	27	Air	760 torr H_2	1010	Winquist et al. (1985a)
	27	10^{-5} torr H_2	10^{-2} torr H_2	184	
	27	Air	760 torr CH_4	490	
Pd–MOS CAP	Room	Air	154 ppm H_2 in N_2	200	Lundström et al. (1986b)

71

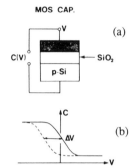

Figure 3.30. (a) Metal–insulator (SiO$_2$)–silicon capacitor sensor. (b) Schematic illustration of the shift of the $C(V)$ curves of the MOS capacitor upon gas exposure (Lundström et al., 1986a).

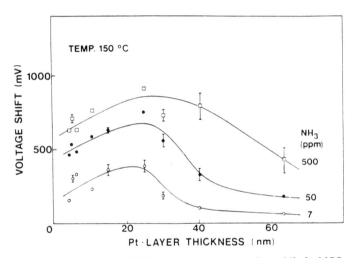

Figure 3.31. Experimentally observed NH$_3$-induced steady state voltage shifts for MOS capacitors with Pt gates vs. the thickness of the Pt layer. Device temperature is 150 °C. Results are shown for three different ammonia concentrations in synthetic air (Lundström et al., 1989).

Sensitivity to other hydrogen-containing gases has been documented by several groups. A significant shift in the $C-V$ curve of the Pd–MOS capacitor sensor following acetylene (C$_2$H$_2$) pulses has been reported (Dannetun et al., 1988a). Hydrogen pulses were introduced before and after acetylene exposures with remarkably similar fast initial shifts, indicative of similar hydrogen coverages on the Pd surface. This behavior was explained by assuming complete dehydrogenation of the acetylene. Similar behaviors have also been observed for many other unsaturated hydrocarbons in the C$_2$–C$_4$ range, leading to the conclusion that the dehydrogenation of these C$_x$H$_y$ gases is

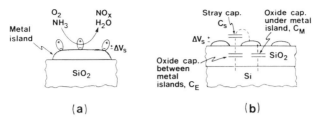

Figure 3.32. Schematic illustration of a recent model for the observed ammonia sensitivity of discontinuous metal films: (a) adsorbed ammonia molecules and/or reaction intermediates give rise to a dipole layer on the metal, i.e., a surface potential change, ΔV_s; (b) the surface potential change is capacitively coupled to the semiconductor surface, giving rise to an equivalent voltage shift of the $C-V$ curve, $\Delta V \leqslant \Delta V_s$ (Lundström et al., 1989).

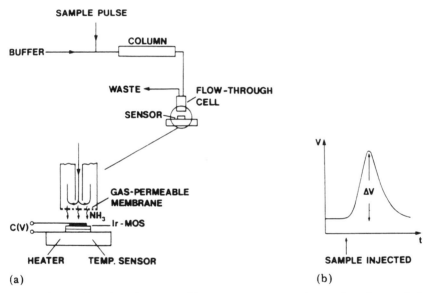

Figure 3.33. (a) Flow-through system with a separate enzyme column where the sample is injected in short pulses. (b) Definition of the pulse response of the ammonia sensor to an injected sample pulse. If the pulse length is short enough, ΔV will be proportional to the NH_4^+ ion concentration of the sample passing behind the gas-permeable membrane (Lundström et al., 1986).

also complete on the Pd surface and independent of the molecular geometry or the carbon–carbon band order (Dannetun et al., 1988a). Table 3.3 presents some details as to the hydrocarbons studied by Dannetun et al. (1988a). Figure 3.34 shows the response of the sensor to acetylene and ethylene, two unsaturated hydrocarbon gases at 500 K. The slow response following the

Table 3.2. Observed Pulse Responses in a Flow Injection System with a Column Containing Immobilized Enzyme[a]

Substrate	Enzyme	Buffer pH	Pulse response (mV)
Ammonia	—	7.7	10
Ammonia	—	12.7	28
Urea	Urease	8.5	16
Creatinine	Creatinine iminohydrolase	8.5	8
Glutamate	Glutamate dehydrogenase	8.5	8
Alanine	Alanine dehydrogenase	9.0	9
Histidine	Histidine ammonialyase	9.0	6
Tryptophan	Tryptophanase	8.5	5
Adenosine monophosphate (AMP)	AMP-deaminase	9.0	10

Source: Lundström et al. (1986a).

[a]Substrate pulse: $10\,\mu M$ for 30 s with a flow rate of 0.125 mL/min. Sensor: thin Ir-gate MOS device at a temperature of $40\,°C$; this device had $\Delta V_{max} \approx 80\text{--}90\,mV$ for ammonia gas pulses 60 s long. The accuracy of the pulse responses was better than 1 mV.

Table 3.3. The Saturation Work Function shifts of Some Unsaturated Hydrocarbons on Polycrystalline Pd at 300 K[a]

Gas	Structure	$\Delta\Phi$ (eV)	$\dfrac{d(\Delta\Phi)}{dt_i}$ (norm.)	Carbon flux (rel. H flux)
Ethylene	C=C	-1.00 ± 0.04	0.48	1/2
Propene	C=C—C	-1.42	0.59	1/2
Butene	C=C—C—C C—C=C—C	-1.46	0.48	1/2
Propadiene	C=C=C	-1.62	0.72	3/4
Butadiene	C=C—C=C	-1.50	0.60	2/3
Acetylene	C≡C	-1.69	1.00	1
Propyne	C≡C—C	-1.72	0.68	3/4
Butyne	C≡C—C—C	-1.73	0.58	2/3

Source: Dannetun et al. (1988a).

[a]The initially applied pressures were set to produce equivalent H flux. The initial rates of the work function shifts normalized to acetylene are also presented.

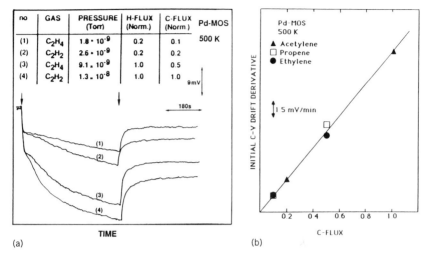

(a)

(b)

Figure 3.34. (a) Comparison of $C-V$ shifts for four different hydrocarbon pulses, with pairwise equal hydrogen fluxes, but with different carbon fluxes. The initial fast shifts are pairwise equal since they only depend on the amount of impinging hydrogen atoms; the dehydrogenation is complete. The slow shift, however, differs for all four pulses, since it depends on the flux of carbon atoms (pressures shown are uncorrected for gauge sensitivity). (b) The initial derivative of the slow, carbon-induced drift vs. carbon flux in the corresponding hydrocarbon exposure (Dannetun et al., 1988a).

rapid initial transient is due to the growing layer of carbon on the surface after dehydrogenation (Lundström et al., 1989). At lower temperatures the response of the sensor to the hydrocarbon pulse becomes much more complicated, since there no longer occurs complete dehydrogenation on the surface. The total dehydrogenation of unsaturated hydrocarbons at $T > 400\,K$ is much more difficult to detect using other surface sensitivity techniques. In any case, other kinds of detection would be more cumbersome and would require sensitive calibration to produce quantitative results, such as spectroscopic probes [secondary ion mass spectrometry (SIMS), X-ray photoelectron spectroscopy (XPS), etc.].

The alkanes (e.g., ethane) are not likely to cause the response of the Pd–MOS capacitor sensor in the $T < 250\,°C$ range, as no dehydrogenation occurs (Lundström et al., 1989). It is interesting to note, however, that methane (CH_4) has been detected via its interaction with an active Pd layer: the $C-V$ curve of the sensor exhibits a shift in 1000 ppm CH_4 in argon or in air (Maclay et al., 1988). Sensor response saturation times for 1000 ppm CH_4 in synthetic air are on the order of 1 min. In vacuo, down to 6×10^{-5}

torr methane pressure has given rise to a signal from the Pd–MOS capacitor sensor (Pietrucha and Lalevič, 1988).

Alcohols can be detected using the capacitor sensor. Ackelid et al. (1987) have measured the response to ethanol (C_2H_5OH) both at high and low pressures. Clean-surface experiments in vacuo have indicated that ethanol does not dehydrogenate at $T > 425$ K. At higher pressures and in an air ambient, however, the dehydrogenation appears to proceed even at higher rates, perhaps promoted by the pressure of oxygen.

The $C–V$ characteristic of the Pd–MOS sensor does show a response to oxygen exposure under ambient conditions [see Eqs. (3.90)–(3.92) and the discussion following these equations]. Under UHV conditions the shift is opposite to what is expected for hydrogen exposure, thus indicating an emptying of hydrogen sites at the Pd–SiO$_2$ interface. The shift is fairly large, ca. 50–100 mV under UHV (Fig. 3.35). With clean surfaces, the dehydrogena-

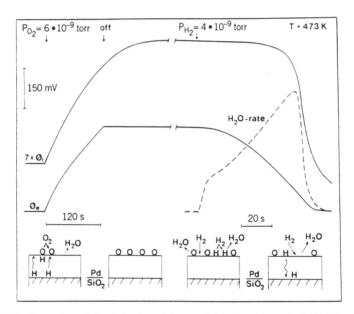

Figure 3.35. Illustration of the behavior of the work function, ϕ_e, and $C–V$ shift, ϕ_i, upon exposure of the Pd–MOS structure to oxygen, followed by hydrogen exposure. The mass spectrometer signal from water ($m/e = 18$) during the hydrogen exposure is also shown. During the oxygen exposure the small amount of residual hydrogen at the interface will be consumed in a water-forming reaction at the surface. This is seen by the negative hydrogen response of the $C–V$ curve. Oxygen will also absorb on the Pd surface, leaving an oxygen layer ($\leqslant 0.25$ monolayer Pd-equivalency), as illustrated by the increase of the work function. When the hydrogen is applied, oxygen will be consumed and desorb as water. Hydrogen does not appear at the interface until almost all oxygen has been consumed (Petersson et al., 1984).

tion of the interface continues long after the oxygen exposure has been shut off. The effect of oxygen-induced surface and interface dehydrogenation of the Pd–MOS capacitive sensor requires, of course, at least a small amount of hydrogen background, which will have no significant effect on the surface chemical reaction. The resulting sensitivity to oxygen has been further utilized to study the dissociation of oxygen-containing molecules such as nitric oxide (NO) and carbon monoxide (CO) (Dannetun, 1987). The latter molecule is known not to dissociate on Pd, a fact witnessed by the absence of any sensor signal upon exposure to CO in the temperature range 353–473 K. At the highest temperature the oxygen and NO responses were found to be similar, indicating that NO dissociates on the Pd surface. At the lowest temperature, the NO response was similar to the CO response, thus indicating that the NO molecule does not dissociate in the 350 K range. Lundström et al. (1989) have given an excellent account of the chemical reaction kinetics of the oxygen–hydrogen system under UHV conditions with Pd or Pt catalysts. Many other interesting catalytic phenomena have been observed in the presence of O_2 by using MIS capacitive sensors under UHV. Among these we should mention the combustion of unsaturated hydrocarbons (C_2–C_4) to H_2O and CO_2 on an oxygen-covered Pd surface at temperatures above 475 K, which results in the complete removal of all oxygen from the surface (Dannetun et al., 1988b).

When ammonia contacts the Pd surface, it does not give rise to a measurable signal in the sensor, because NH_3 does not dissociate on the bare Pd surface. In contrast to this behavior, oxygen-covered Pd surfaces promote the dissociation of NH_3 molecules. This produces water molecules in the vicinity of the oxygen atoms on the Pd surface, with a maximum desorption rate early in the process when the oxygen coverage is large (Fogelberg et al., 1987; Petersson et al., 1987a, b). In this case the low-concentration background hydrogen eventually fills all the emptied Pd–SiO_2 interface sites after all the oxygen has been consumed through H_2O desorption, and the sensor acts as a monitor of the oxygen coverage on the Pd surface. Similar phenomena have been observed with ethanol, which combusts to CO_2 and H_2O on oxygen-covered Pd under UHV conditions (Ackelid et al., 1987).

Very recently, Lundström et al. (1990) tested three types of MIS sensors in the presence of carrier gas flows of 20% O_2 in N_2 or Ar, in which they injected several test gases at appropriate flow rates. The devices tested were both of the capacitor and the transistor type and gave similar information. They were (a) Pd–MOS, with a thick ($\geqslant$ 100 nm) Pd-gate; (b) Pt/Pd–MOS, with a gate consisting of 15 nm of Pt on top of thick Pd; and (c) Pt–TMOS and Ir–TMOS, with thin layers (3–10 nm) of the catalytic metals as the active gates. It was observed that at a given operating temperature, a thin layer of Pt on top of a Pd-gate increases the sensitivity to hydrogen decreases. This

Table 3.4. The Response of a Pt–TMOSFET, Operated
at 190 °C, for Various Compounds[a]

Compound	Sensor Response (mV)
Ammonia	195
Ethanol	147
Acetone	97
Ethylene	92
Acetic acid	64
Ethyl acetate	51
Toluene	41
Cyclohexane	23
Methane	0

Source: Lundström et al. (1990).
[a]The concentration of the compound was 100 ppm, and it was subjected to the sensor for 30 s.

occurs presumably owing to a change in catalytic activity of the metal surface. Hughes et al. (1987) have made similar observations by alloying the Pd-gate with Ag.

Table 3.4 shows the results of a screening test of several compounds exposed to a Pt–TMOSFET at 190 °C. The sensitivity was found to be temperature dependent, with the temperature dependence arising from the dehydrogenation reaction on the oxygen-covered metal surface. The thin metal gates (Pt or Ir) have exhibited sensitivity to the NH_3–H_2 water vapor mixtures of gases, with the response caused by NH_3 being about the same with and without the saturated water vapor ambient for the Ir electrode. As Table 3.5 shows, the respective response to NH_3 of the thin Pt gate was considerably smaller in the presence of water vapor. This table further illustrates another interesting difference between Pt and Ir. For Pt the voltage shift in an NH_3–H_2 mixture is larger than the sum of the NH_3 and H_2 shifts, implying chemical synergy on the Pt surface in the NH_3–H_2 mixture. The opposite is seen for Ir, which implies that competing or blocking reactions take place on the Ir surface in the mixture. The possibility of utilizing MOS-type gas sensors for the monitoring of mixtures of gases, based on different detection mechanisms, is currently under intense investigation using pattern recognition principles (see Section 3.5.3.3, below). Figure 3.36 is an illustration of the proposed detection mechanism for dual mixtures such as ethylene and ethanol. This schematic complements Fig. 3.32, an earlier capacitive coupling

Table 3.5. The Sensitivity, ΔV, of NH_3/H_2 and NH_3/H_2O Compared with the Mixture of the Gases[a]

	Sensitivity, ΔV (mV)	
Gas	Pt	Ir
NH_3	625	620
H_2	100	125
$NH_3 + H_2$	815	645
(due to NH_3)	(715)	(520)
NH_3	500	550
H_2O	165	125
$NH_3 + H_2O$	545	645
(due to NH_3)	(380)	(520)

Source: Lundström et al. (1990).

[a]Operating temperature: 150 °C. Gas concentrations: NH_3, 100 ppm; H_2, 500 ppm; H_2O, 100% RH.

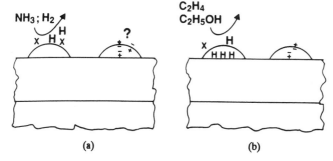

Figure 3.36. Schematic illustration of the different detection mechanisms for the catalytic gate metal MOS sensors. (a) For simultaneous exposure to, e.g., ammonia and hydrogen, the voltage shift arises from both the metal–insulator interface and the surface of the metal. The possibility of "negative" dipoles, due to reactions between hydrogen and ammonia on the surface, is indicated. (b) From ethylene or ethanol, hydrogen will be dissociated and the shift arises from the metal–insulator interface, but possibly also from molecules or reaction intermediates on the metal surface. In all cases dipoles/charges between the metal islands may also contribute to the observed voltage shifts (Lundström et al., 1990).

model for the ammonia sensitivity of thin metal films (Lundström et al., 1986b).

A very-high-temperature (VHT) MOS capacitor sensor has been fabricated using a Pd–MOS sensor together with a Pt coil catalyst (Hornik, 1990). The Pt coil is mounted together with the sensor on one holder, so that the temperature of the coil can be increased to more than 1100 °C, while the Pd–MOS sensor itself is kept at a constant temperature between 100 and 190 °C. The idea behind the use of VHT hydrogen sensors stems from the fact that the conventional Pd–MOS devices, besides their sensitivity to hydrogen itself, exhibit sensitivity to hydrogen-containing molecules like ethanol, hydrogen sulfide, ammonia, and some hydrocarbons (Zhang and Zhao, 1988; Fare et al., 1988; Ackelid et al., 1986), caused by thermally activated dissociation of these molecules at the Pd surface. This leads to hydrogen generation, with a maximum operating temperature of the silicon-based device being limited to ca. 250 °C. The presence of a mounted Pt catalyst, the temperature of which can exceed 1100 °C, allows the dissociation of several other molecules, notably trichloroethylene, acetone, isobutane, n-hexane, chloroform, and ether. The sensor structure is shown in Fig. 3.37. The high-temperature catalyst is operated in a Wheatstone bridge, together with a control circuit that keeps its resistance—and therefore its temperature—constant. Using the theoretical response of the sensor in H_2 [Eqs. (3.87) and (3.89)], Hornik showed that the equivalent hydrogen partial pressure at the catalyst is proportional to the test gas consumption:

$$p(H_2)_e = \frac{X_i A C \exp(-E_a/kT)}{1 + R_d C \exp(-E_a/kT)} \tag{3.102}$$

where X_i is the initial gas concentration; A and C are proportionality

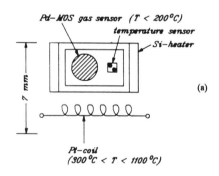

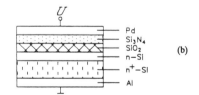

Figure 3.37. Hybrid structure with Pd–MOS gas sensors and Pt coil: (a) top view; (b) cross-sectional view and high-frequency C–V characteristics of the MOS capacitor (Hornik, 1990).

constants; R_d is a "resistive" conversion limiter; and E_a, the gas-specific activation energy. In n-hexane at $T_{sensor} = 190\,°C$, good agreement of the sensor signal with Eq. (3.102) as a function of the Pt coil temperature T was found for a 100 ppm sample, down to ca. 650 °C. Deviations were observed at lower temperatures, because there is always some residual hydrogen in the Pd layer keeping the capacitive voltage at or above a minimum value ΔV_{min}.

3.5.2.2. Device Fabrication and Performance

Regarded as a microelectronic system, the MOS capacitor using a catalytic metal electrode can be represented by the following operating principle on Pd film (D'Amico and Verona, 1988): catalytic reaction diffusion interface process of H through the Pd—

$$H_2 \rightarrow H + H \rightarrow \Delta Q \quad \text{(adsorption)}$$

$$O_2 \rightarrow OH, \quad H_2O \rightarrow -\Delta Q \quad \text{(desorption)}$$

The conventional method for investigating the adsorption–absorption–desorption processes with an equivalent charge $\pm \Delta Q$ injected to $(+)$ or extracted from $(-)$ the Pd surface is the $C-V$ measurement involving the shift in the flat-band voltage [Eq. (3.48)]. In practice the shift of the $C-V$ curves under adsorption or desorption is measured as ΔV_{FB} and is then correlated to the hydrogen concentration. Usually, these measurements are carried out at elevated temperatures (ca. 150 °C), which require suitable electronic circuits to allow the conversion between the $C-V$ measurements and the output voltage either in analog or digital form vs. the H_2 concentration.

As described in Section 3.1, a conventional MOS capacitor is easy to fabricate using silicon substrate, growing a thermal oxide, and evaporating a thin metal electrode. For a gas-sensitive device, Pd is evaporated on the front side of the wafer through a mechanical mask, thus producing Pd electrodes of suitable geometry [e.g., dots of 1 mm^2 area (Lundström and Svensson, 1985)]. Palladium is known to have very weak adhesion to SiO_2, which causes the Pd film to peel off during fabrication (Shivaraman and Svensson, 1976) or form blisters during use (Armgarth and Nylander, 1982). Heat treatment in air at 200 °C and the use of Pd films thicker than 400 nm have been shown to bypass these problems. Another solution can be the use of an intermediate chromium layer between the palladium and the oxide, which nevertheless changes the sensitivity of the device. A complete sensor is made by combining the capacitor with a heater and a temperature sensor to keep the device at an appropriate constant temperature (Lundström and Svensson, 1985). A temperature regulator as well as a sensitive capacitance meter are needed to measure the sensor capacitance as a function of the gas concentration at constant bias (Steele et al., 1976) or to keep the capacitance constant via a regulator and vary the bias voltage as a function of gas concentration (Lundström et al., 1977).

In practical devices integration is favored at the wafer level. Such fabrication sequence follows standard electronic device processing. An example is shown in Fig. 3.38, which is a layout of a capacitor gas sensor circuit with its peripherals (Poteat and Lalevič, 1982). For the fabrication of special metal–nitride–oxide–semiconductor (MNOS) device structures, Si_3N_4 dielectric layers were deposited by thermal decomposition of silane (SiH_4) and ammonia (NH_3) on 500 Å SiO_2 layers grown on 4–$10\,\Omega$·cm, p-type $Si\langle 100\rangle$ or 6–$12\,\Omega$·cm, n-type $Si\langle 100\rangle$ in a dry oxygen ambient at $1000\,°C$. Transition metal gates (Pd, Pt, and Ni) were deposited through a shadow mask of area $4.42\times 10^{-3}\,cm^2$ to form the active capacitor elements. Figure 3.38(b) shows the test system.

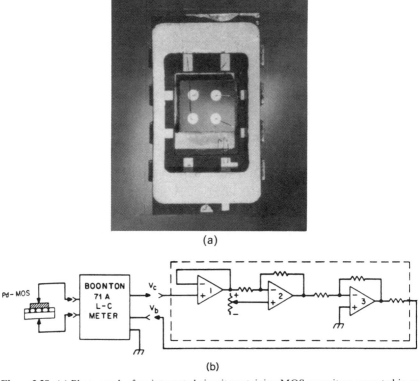

(a)

(b)

Figure 3.38. (a) Photograph of an integrated circuit containing MOS capacitors mounted in an 8-pin package. (b) Circuit for testing in the constant capacitance mode (Poteat and Lalevič, 1982).

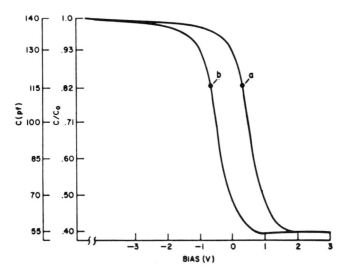

Figure 3.39. $C-V$ characteristic for p-type MOS capacitor with Pd gate prior to (curve a) and after (curve b) absorption of H_2 at 760 torr (1 MHz signal frequency). (Poteat and Lalević, 1982).

MOS capacitor sensitivities were tested in the H_2 partial pressure range of 10^{-8} to 760 torr. Figure 3.39 shows typical $C-V$ characteristics of the test system. Table 3.6 details the performance of MOS and MNOS capacitors. In fabricating optimized MOS capacitor sensors, all normal MOS drift phenomena must be minimized. This requires highest quality SiO_2 growth, resulting in alkali-ion- and radiation-damage-free layers (Evans et al., 1986; Choi et al., 1986; Dobos et al., 1984; Fare and Zemel, 1987). Similarly, the reliability of Pd–MOS capacitive sensors can be improved by ensuring that Pd surface impurities such as carbon, sulfur, and oxygen are absent. All these molecules interfere with the kinetic process of gas permeation through the Pd thin film (Lalauze et al., 1988). Criteria for the acceptable range of sensitivities to H_2 of the Pd-gate MOS sensor have been discussed by Lundström and Svensson (1985).

3.5.3. MIS Field Effect Transistor (FET) Sensors

3.5.3.1. Devices, Mechanisms, and Applications

The MOS transistor theory presented in Section 3.3.2, Eqs. (3.47)–(3.52), yields simple relations for the drain current, I_D, in the unsaturated region, i.e., for $V_D < V_G - V_T$. Specifically, Eq. (3.50) can be simplified (Bergveld,

Table 3.6. Summary of Flat-Band Voltage Changes and Absorption Time Constants in MOS and MNOS(∗) Capacitors Due to Absorption of H_2, CH_4, C_4H_{10}, and CO Gases

Gate Metal	Dielectric	Gas Type	Gas Pressure (mmHg)	ΔV_{FB} (mV)	Absorption Time constant (s)
Pd	SiO_2	H_2	2×10^{-8}	215	1.2×10^4
Pd	SiO_2	H_2	4×10^{-3}	255	6.5×10^2
Pd	SiO_2	H_2	2×10^{-2}	285	5.0×10^2
Pd	SiO_2	H_2	0.5	560	2.0×10^2
Pd	SiO_2	H_2	1.0	775	1.8×10^2
Pd	SiO_2	H_2	760	1240	19×10^{-2}
Pd(∗)	Si_3N_4	H_2	2×10^{-2}	285	5.0×10^2
Pd(∗)	Si_3N_4	H_2	0.1	420	3.5×10^2
Pd(∗)	Si_3N_4	H_2	760	1210	19×10^{-2}
Pd	SiO_2	CH_4	760	490	1×10^3
Pd	SiO_2	C_4H_{10}	760	750	5.8×10^2
Pd	SiO_2	CO	760	100	2.6×10^4
Pt	SiO_2	H_2	760	610	5.5
Pt	SiO_2	CH_4	760	—[a]	—[a]
Ni	SiO_2	H_2	760	120	$> 10^4$
Ni	SiO_2	CO	760	140	1.4×10^4

Source: Poteat and Lalevič (1982).
[a] No response.

1985):

$$I_D = \mu C_{ox}(W/L)[(V_G - V_T)V_D - \tfrac{1}{2}V_D^2] \qquad (3.103)$$

where μ is the electron mobility in the channel; $C_{ox}(\equiv C_i)$ is the oxide capacitance per unit area; W/L is the channel width-to-length ratio; and the rest of the symbols have the assigned definitions in Section 3.3.2 [V_G and V_D are the applied gate-source and drain-source voltages, respectively, and V_T is the threshold voltage; see Eq. (3.48)]. In the saturated region Eq. (3.103) must be replaced by

$$I_D = \tfrac{1}{2}\mu C_{ox}(W/L)(V_G - V_T)^2 \qquad (3.104)$$

Note that V_T depends on the flat-band voltage, V_{FB} [Eq. (3.48)], which is a function of the gate metal work function, ϕ_m [Eq. (3.47)]. When a catalytic metal such as Pd is deposited instead of the aluminum commonly used for gates in MOS technology, a hydrogen sensor results (Lundström et al., 1975b). The sensitivity of this device has been properly described by the effect of the ambient hydrogen gas on the gate metal work function (Bergveld, 1985):

$$\Delta\phi_m = (\Delta\phi_{m,\max}) \left\{ \frac{K[p(H_2)]^{1/2}}{1 + K[p(H_2)]^{1/2}} \right\} \tag{3.105}$$

where K is an overall equilibrium constant, and $\Delta\phi_{m,\max}$ is the maximum shift of the metal work function, which occurs at total coverage with H atoms of the available number of adsorption sites per unit area at the Pd/SiO_2 interface.

The operating temperature of MOSFET gas sensors is usually elevated (50–150 °C) in order to speed up the catalytic reactions on the Pd surface. The response time upon H_2 introduction is on the order of 5 s for 50 ppm hydrogen in air at 150 °C, and it decreases with temperature with an activation energy of ca. 0.3 eV (Lundström et al., 1977).

The general principle of a contact potential shift between the catalytic gate metal and the substrate silicon in the presence of an ambient gas interacting with the metal has been known to be valid also with other gases such as CO. In this case the Pd gate should be porous (Krey et al., 1983). A large ammonia sensitivity can be obtained by using thin catalytic metal films as gates (Winquist et al., 1985b).

The majority of MOSFET gas sensors differ in the details of the fabrication of the gate areas, particularly the gate metal. Following Bergveld (1985), a classification of devices can be made:

(i) The Open-Gate Field Effect Transistor (OGFET). The OGFET was introduced by A. Johannessen in 1970 and further studied by Thorstensen (1981). This gas sensor is a MOSFET without the gate metal, with a 2–50 nm oxide exposed to a gaseous environment. Responses to polar and nonpolar gases have been obtained with a strong variation of the drain current as a function of the partial pressure of the polar gas (e.g., water or methanol). Figure 3.40 shows a schematic of the OGFET cross section. Even though the operating mechanism has not been well established, it has been surmised that dissociated gas molecules at the surface are transported along the oxide surface under the influence of the electric fringe field resulting from the drain-source voltage. The diffusion of charged and uncharged particles into the oxide toward the $Si–SiO_2$ interface may also play a role. A change in the dielectric constant of the oxide has been further associated with drain current modulation. These various possibilities of operation are shown symbolically

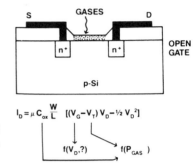

Figure 3.40. Schematic representation and proposed operation mechanism of the OGFET (Bergveld, 1985).

in Fig. 3.3. No substantial development of the OGFET has taken place, most likely due to our incomplete understanding of this device.

(ii) The Adsorption Field Effect Transistor (ADFET). A version of the OGFET utilizing ultrathin gate oxide thickness ($\leqslant 50\,\text{Å}$) has been given the name ADFET (Cox, 1974). For oxide thicknesses in this range only, the ADFET appears to respond to gases exhibiting a permanent net dipole moment, such as H_2O, NH_3, HCl, CO, NO, NO_2, and SO_2. It is possible that the drain current is determined by the fringing electric field of the adsorbed molecules. The device structure and operation are summarized in Fig. 3.41. Any selectivity of the device was claimed to occur as the result of replacing the silicon–oxygen bonds by silicon–carbon bonds, which are in turn modified with special groups. The obvious disadvantage of both the OGFET and ADFET sensors is the open-gate construction, which renders

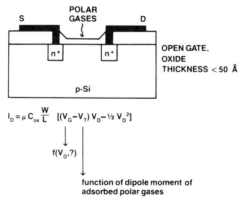

Figure 3.41. Schematic representation and proposed operation mechanism of the ADFET (Bergveld, 1985).

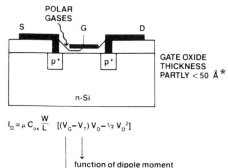

$$I_D = \mu\, C_{ox}\, \frac{W}{L}\; [(V_G - V_T)\, V_D - \tfrac{1}{2}\, V_D{}^2]$$

function of dipole moment
adsorbed polar gases

controlling electrical and
chemical operation

Figure 3.42. Schematic representation and proposed operation mechanism of the SAFET (Bergveld, 1985). *The gate insulator consists partly of an airgap created by underetching the polysilicon gate (Stenberg and Dahlenbäck, 1983).

them sensitive to electric interference sources that cause an electric field in the oxide. For this reason their popularity has not been very high.

(iii) The Surface-Accessible Field Effect Transistor (SAFET). In the SAFET the gate insulator consists partly of an airgap created by underetching the polysilicon gate (Stenberg and Dahlenbäck, 1983). In this manner the gas has direct access to the silicon surface without the loss of the electrostatic shielding action of the gate metal, thus offsetting the disadvantage of the OGFET and the ADFET devices. Polar gases are known to increase the drain current of the SAFET. The operating mechanism is expected to be similar to ADFET, assuming that the underetched gate region is covered with a native oxide. A schematic representation is shown in Fig. 3.42. The existence of the gate metal appears to be very useful toward the improvement of the reversibility of the adsorption process, the sensitivity and the stability of the device.

(iv) The Charge-Flow Transistor (CFT). The CFT can be regarded as a modification of the MOSFET, where the gate does not cover the actual gate area but is placed beside it or it circumscribes it (Senturia et al., 1977). The gate area itself is covered with a resistive material in contact with the laterally shifted gate. A time delay appears between the onset of a voltage step between the eccentric gate and the source and the appearance of a complete channel. This delay is partly a function of the film resistivity, which in turn depends on environmental conditions such as humidity (Fig. 3.43). In fact, the MOS fabrication of this device offers the possibility of measuring the resistance of very-high-resistivity thin films more accurately and easily than more conventional techniques.

(v) The Pressure Field Effect Transistor (PRESSFET). If the material

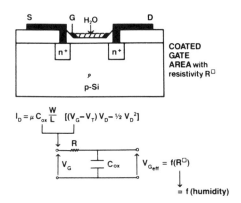

Figure 3.43. Schematic representation and basic operation of the CFT (Bergveld, 1985).

between the elevated gate of Fig. 3.43 and the gate oxide is a dielectric, rather than the resistive material used with the CFT, the device will still operate as a MOS transistor. As a result of this change the oxide capacitance per unit area, C_{ox}, in Eq. (3.103) must be replaced by C_{eq}, the equivalent capacitance of the dielectric sandwich on top of the gate; C_{eq} is sensitive to environmental factors such as an external pressure changing the distance between the original gate oxide and the elevated gate. The role of the dielectric may be assumed by air or vacuum. Bergveld (1985) has given the obvious modifications of the drain current for the PRESSFET (Fig. 3.44):

$$I_D = \mu C_{\text{airgap}} \left(\frac{W}{L}\right)\left[\left(V_G - 2\phi_F - \phi_{ms} + \frac{Q_{\text{tot}}}{C_{\text{airgap}}}\right)V_D - \frac{1}{2}V_D^2\right] \qquad (3.106)$$

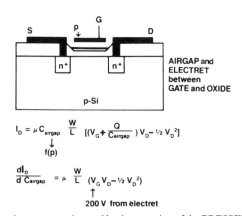

Figure 3.44. Schematic representation and basic operation of the PRESSFET (Bergveld, 1985).

where $C_{eq} \approx C_{airgap} \ll C_{ox}$; ϕ_F is the Fermi potential difference between the doped bulk silicon and intrinsic; ϕ_{ms} is the contact potential (the difference between metal and silicon work functions); and $Q_{tot} \equiv Q_B + Q_{it} + Q_F$ (Q_B being the bulk depletion charge per unit area; Q_{it} and Q_f, the charge per unit area of the interface traps and the fixed oxide charged, respectively). The sensitivity of the device to a variation of the airgap capacitance is

$$\frac{dI_D}{dC_{airgap}} = \mu\left(\frac{W}{L}\right)\left[(V_G - 2\phi_F - \phi_{ms})V_D - \frac{1}{2}V_D^2\right] \qquad (3.107)$$

Large external applied gate-source voltages, V_G, increase the sensitivity of this device. A particularly successful design includes the piezoelectric poly(vinylidene fluoride) (PVDF) as a dielectric layer between the metal gate and the oxide of the PRESSFET. External biasing results in the well-known electret microphones, in which a Teflon layer may be inserted in the airgap, resulting in an equivalent V_G on the order of several hundred volts.

Although the PRESSFET concept has not been directly associated with gas detection, its application in the area of gas-phase photoacoustic pressure sensing has been increasing in recent years (Sigrist, 1992; see Chapter 4).

(vi) The Suspended-Gate Field Effect Transistor (SGFET). A device with a separated gate electrode and a SiO_2/Si_3N_4 gate insulator of a "normal" thickness is the SGFET (Blackburn et al., 1983). In the SGFET, a suspended platinum mesh is used as the gate electrode and the adsorption of polar gases on the inner platinum surface modifies the gate field and thus the drain current. On this platinum mesh, a chemically sensitive layer can be deposited. In an application, a SGFET with a layer of electrochemically deposited polypyrole on platinum exhibited a response to lower aliphatic alcohols (Josowicz and Janata, 1986).

The first MOSFET device for hydrogen detection was reported by Lundström et al. (1975a). Lundström's group in Sweden has played a leading role in the development of Pd-gate MISFET gas sensors. According to Lundström (1981), when the device is exposed to hydrogen, dissociated hydrogen atoms absorbed at the Pd–SiO$_2$ interface are polarized and give rise to a dipole layer. This corresponds to a voltage drop, which is added to the externally applied gate voltage, V_G. The characteristics of the sensor consist of this voltage shift by ΔV. A typical cross-sectional view and voltage shift of a Pd-MOS transistor are shown in Fig. 3.45. The actual change in work function is assumed to be proportional to the interface concentration of absorbed hydrogen, or the hydrogen coverage. The maximum change occurs when the hydrogen coverage is 1, i.e., each interface site is occupied by one hydrogen atom. The maximum change, ΔV_{max}, has been found (by extrapolation) to be ca. 0.5 V (Lundström and Svensson, 1985).

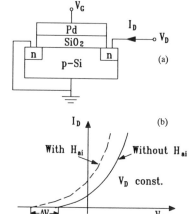

Figure 3.45. (a) Cross-sectional view and (b) current–voltage characteristics of Pd-MOSFET. Here D = drain; G = gate; H_{ai} = hydrogen concentration at the Pd-oxide interface (Lundström, 1981; Lundström et al., 1989).

The large sensitivity normally observed for MOSFET sensors such as the one shown in Fig. 3.45 is largely due to the fact that such sensors respond to surface adsorbates in a Langmuirian fashion (Lundström and Svensson, 1985). Schematically, the chemical reactions on the metal surface may be described by (Lundström et al., 1986a)

$$\frac{d\theta}{dt} = c_1 P(1 - \theta)^\alpha - c_2 \theta^\beta \qquad (3.108)$$

where P is the partial pressure of the detected species and θ is its surface coverage; c_1 and c_2 are temperature-dependent rate constants, which may also depend on the ambient (e.g. the oxygen pressure). For the hydrogen sensor in the presence of oxygen, $\theta = \theta_i$, the interfacial coverage, $\alpha = \beta = 2$, and c_2 is a function f of $p(O_2)$, so that Eq. (3.108) becomes at steady state

$$\frac{\theta_i}{1 - \theta_i} = K_1 \left\{ \frac{c_1 p(H_2)}{f[p(O_2)]} \right\}^{1/2} \qquad (3.109)$$

MOSFET gas sensors with catalytic metal gates are often operated at an elevated temperature to speed up the response and to avoid difficulties with adsorbed water molecules. Under these conditions the device gives a fast and reversible response to hydrogen. Its sensitivity depends on the oxygen pressure and sometimes also on the nitrogen pressure. Low-temperature operation is also possible. There appear to be two main differences between room-temperature and high-temperature operation (Lundström and Söderberg,

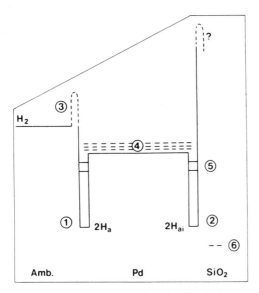

Figure 3.46. Schematic summary of the adsorption sites for hydrogen and other observed phenomena in the Pd–SiO$_2$ system: (1 & 2) "normal" adsorption sites at the surface and interface, respectively, with similar properties; (3) an extra barrier for the dissociation of hydrogen, probably related to adsorbed oxygen atoms; (4) bulk adsorption sites filled to any extent only in an inert atmosphere (at low temperatures and/or high hydrogen concentrations); (5) hydrogen adsorption sites with small adsorption energies filled in experiments with hydrogen in an inert atmosphere and temperatures around 100–150 °C; (6) hydrogen introduced in the oxide probably interacting with (sodium) ions in the oxide (Lundström and Svensson, 1985).

1981/82): room-temperature operation is limited to smaller device sensitivities and longer response time constants. For nonporous films the exact mode of operation at low temperatures is not known. The deep hydrogen adsorption sites, with adsorption energies of ca. 1 eV/H$_2$, will be saturated at extremely low hydrogen concentrations, especially since water adsorption on the Pd surface appears to block the reaction between hydrogen and oxygen. Lundström and Svensson (1985) have suggested that at room temperature other adsorption sites or even bulk hydrogen give rise to the observed signals. Figure 3.46 schematically displays a summary of the known energetics of the H$_2$–Pd–SiO$_2$ system. For porous metal films the gas–surface interaction pathways are more complicated than is the case with nonporous films (Yamamoto et al., 1980; Morrison, 1987).

In general, a practical sensor chip consists of the gas-sensing MOSFET, an integrated heater (resistor), and a temperature sensor (diode), as shown in Fig. 3.47. Apart from the normal problems in semiconductor processing, the problem of Pd adhesion to the SiO$_2$ is a particularly challenging one

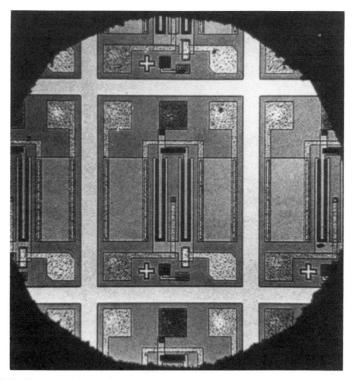

Figure 3.47. Photograph of an integrated circuit containing heating resistors (the two outer rectangles), a diode for temperature control (in the middle), and a Pd-gate MOS transistor (in two parts around the middle). The transistor gate is shorted to its drain (upper middle contact). The chip size is about 0.5×0.7 mm (Lundström and Svensson, 1985).

(Choi et al., 1984). The temperature of the sensor can change its selectivity toward certain gases. In experiments with hydrogen in oxygen in the concentration region of 1–1000 ppm it has been found that Eq. (3.109) fits the data well when precautions are taken to define the true baseline. Figure 3.48 shows an example of the valuation of experimental results for a $Pd-Al_2O_3-SiO_2-Si$ structure and the sensor selectivity change at two different temperatures (Armgarth et al., 1985). The experimental points represent several isotherms of different oxygen and hydrogen pressures plotted vs. a normalized pressure. It can be seen that the data fit Eq. (3.109), with $f[p(O_2)] = [p(O_2)]^{1/2}$ at $T = 75\,°C$ and $f[p(O_2)] = [p(O_2)]^{1/4}$ at $T = 50\,°C$. In the chemical reaction scheme (Lundström and Söderberg, 1981/82)

$$O_2 \underset{d_4}{\overset{c_4}{\rightleftharpoons}} 2O_a \qquad\qquad (3.110)$$

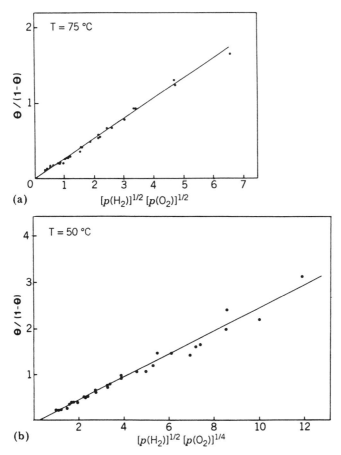

Figure 3.48. Experimental results obtained on a $Pd-Al_2O_3-SiO_2-Si$ structure at (a) 75 °C and (b) 50 °C in different H_2/O_2 mixtures: $\theta = \Delta V/\Delta V_{max}$, where ΔV is the experimentally observed voltage shift and $\Delta V_{max} = 571\,mV$. Note that the experimental points represent a number of isotherms obtained for oxygen concentrations between 10% and 100% and hydrogen concentrations between 0 and 500 ppm. The abscissas were constructed using $p(H_2)$ in ppm and $p(O_2)$ in percent (Lundström et al., 1986a).

$$H_a + O_a \underset{d_5}{\overset{c_5}{\rightleftharpoons}} (OH)_a \tag{3.111}$$

$$(OH)_a + H_a \xrightarrow{c_6} H_2O \quad (gas) \tag{3.112}$$

Several possible models can be considered, depending on the assumptions about the adsorption sites and the relative values of the rate constants. If

$c_6 \ll (c_5, d_5, d_4)$, i.e., if water production is slow, assuming that oxygen is adsorbed on special sites (N_0) that do not compete for hydrogen adsorption, it is found that

$$\frac{n_i}{N_i - n_i} = \left(\frac{c_e d_i}{d_e c_i}\right) \frac{\sqrt{d_5 c_1 p(H_2)\{1 + [c_4 p(O_2)/d_r]^{1/2}\}}}{2 c_5 c_6 N_0 [c_4 p(O_2)/d_4]^{1/2}} \tag{3.113}$$

where rate constants c_1, c_e, d_i, d_e, and c_i are defined in Eqs. (3.84) and (3.85); and $\theta \equiv n_i/N_i$ is the coverage of hydrogen atoms at the surface. The experimental data of Fig. 3.48 represent potentially one special case of the above scheme. It is interesting to note that a change in the oxygen dependence with temperature from $[p(O_2)]^{-1/4}$ to $[p(O_2)]^{-1/2}$ has been observed on Pd–MOS structure by Hua et al. (1984). Experimentally, oxygen dependences of the form $[p(H_2)]^{1/2}/[p(O_2)]^\alpha$ with $1/4 \leqslant \alpha \leqslant 1/2$ have been found.

The alternative scheme

$$O_2 + 2H_a \xrightarrow{c_2} 2(OH)_a \tag{3.114}$$

$$2H_a + 2(OH)_a \xrightarrow{c_3} 2H_2O \quad \text{(gas)} \tag{3.115}$$

yields in the steady state

$$\frac{n_i}{N_i - n_i} = \left(\frac{c_e d_i}{d_e c_i}\right)\left[\frac{c_1 p(H_2)}{2 c_2 p(O_2)}\right]^{1/2} \tag{3.116}$$

in phenomenological agreement with some of the experimental observations. Extensive data of the dependence of the Pd–MOSFET signal on the hydrogen partial pressure in air at several temperature have been presented by Lundström et al. (1977).

These same authors have also presented an absolute reaction rate theory based on principles of heterogeneous catalysis (Thomas and Thomas, 1967) to quantify the rate constants of the proposed gas–surface chemical reactions. Oxygen deactivation of the Pd–MOSFET has been observed by Lundström and Söderberg (1981/82). If the sensor is kept in air or in oxygen without hydrogen for a long time, it responds very slowly the first time it is exposed to hydrogen again. After a few hydrogen exposure cycles, however, it recovers its normal response. It has further been observed that if the device is stored hot in normal air, the background hydrogen in the air keeps the surface clean enough to yield a reasonable response time.

Based on the hypothesis that the MOSFET sensor's response to hydrogen is Langmuirian [see Eq. (3.105)], Lundström and Svensson (1985) calculated the relative differential sensitivity $p[d(\Delta V)/dp]$ and showed that this corresponds to a relative pressure range of $1:10,000$.

In terms of device stability, the most important adverse phenomenon in Pd-MOSFETs has been a hysteresis, or drift, in the hydrogen response, owing to adsorption sites for hydrogen in the oxide at the metal–oxide interface (Nylander et al., 1984). The effect of hydrogen in the oxide can be eliminated by using an insulator other than SiO_2 in contact with the metal (Armgarth and Nylander, 1981; Dobos et al., 1984; Armgarth et al., 1985). A further improvement of the MOSFET sensor has been the elimination of the slow hydrogen adsorption sites in the insulator by introducing more complex structures such as $Pd–Al_2O_3–SiO_2–Si$ devices. The response of these devices to hydrogen is faster, and they can be operated at low temperatures (Armgarth et al., 1985). The origin of the drift problem of Pd–MOSFETs lies in the incorporation of electrical defects in the $Si–SiO_2$ system resulting from the oxidation of single crystalline silicon. These defects include mobile ionic charges, radiation-induced charges, fixed oxide charges, and interface traps (Deal, 1974). One of the most detrimental defects is caused by the alkali ions sodium and potassium. Ion contamination reduction methods have been developed, yet the role of certain defects has not been elucidated yet. Polarization and charge trapping phenomena in the oxide are known to cause a change in the device threshold voltage. The large drift in this quantity is shown on Fig. 3.49 for a p-channel MOS transistor. Most of the drift is due to the so-called negative bias stress instability (Jeppson and Svensson, 1977) at the $Si–SiO_2$ interface, a phenomenon often referred to as "hole trapping." It is possible to decrease the charge-trapping drift phenomena by decreasing the oxide thickness (Lundström, 1981).

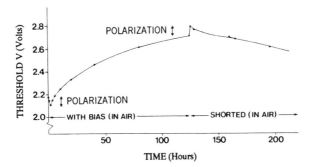

Figure 3.49. Long-term drift in the threshold voltage observed on p-channel Pd–MOS transistors (100 nm oxide). The polarization phenomena are related to charge movement in the oxide, whereas the other instability is due to charge injection. Temperature: 150 °C (Lundström, 1981).

In the case that both oxygen and hydrogen gases are present in a flowing stream, Lundström and Söderberg (1981/82) have observed a kind of hysteresis effect when the oxygen-induced voltage shift was plotted vs. oxygen pressure in the range 0.4–1.4% O_2 in 270 ppm H_2 background at $T = 152\,^\circ C$. The hysteresis appeared as a "turn-on" phenomenon but was absent or negligible at low H_2 background concentrations (e.g., 10 ppm). This and other anomalies in the behavior of Pd–MOSFET sensors have been explored in detail by Lundström and Söderberg (1981/82) in describing the sensitivity to hydrogen and the long-term performance of these devices. More complete explanations can be given with combined additional information supplied by spectroscopic measurements, such as XPS, AES (Auger electron spectroscopy) and FTIR (Fourier transform infrared spectroscopy) under differential device structures, e.g., Pd/Pt-gate and Ag/Cu/Pd/Pt-gate differential transistors (Choi et al., 1984, 1986). The observed signal drift was associated with the decrease of hydrogen adsorption sites on the Pd-gate metal surface, which is oxidized at high temperatures under air ambient operating conditions. The oxidation was observed with XPS and AES. The variation in the sensitivity with time of exposure was shown to be stabilized by high-temperature annealing. Similar annealing effects in the combination of SiO_2–Si_3N_4 double-insulator geometries were reported for the Pd-gate MOSFET H_2 and CO sensors, which exhibit pre-anneal long-term instability (Dobos et al., 1983).

Fundamental understanding of Pd–MOSFET instabilities has centered on the existence different "hydrogen-induced drift" (HID) sites in the oxide with nonuniform heats of adsorption (Lundström et al., 1989). The energy structure of the hydrogen sites on the metal and the oxide sides of the interface must be elucidated before further progress can be made on this issue.

In a recent study Fortunato et al. (1989) used bulk reflectivity and conductance modulation techniques to associate stability issues with a movement of the Fermi level in a Pd film. The data were interpreted in terms of a hydrogen energy level model in the Pd–SiO_2 system with two different energies and sites (near the ambient–Pd and the Pd–SiO_2 interfaces, respectively). Armgarth et al. (1984) have investigated oxide surface charging in Pd–MOS transistors by measuring transient currents induced by changes in bias, and microscopic ellipsometry on the oxide surface. They concluded that surface charges are protons, which is consistent with a "memory effect" of earlier oxide surface charging based on the ability of the Pd gate to store hydrogen easily. Mokwa et al. (1987) have succeeded in eliminating the hydrogen-induced drift from n-channel MOS transistors by use of NMOS metal-gate technology (Zimmer et al., 1979) and thus have fabricated a sensitive Pd-gate NMOS device with 40 nm Si_3N_4 on top of 40 nm SiO_2, capable of detecting arsine (AsH_3) concentrations of 0.1–30 ppm mixed with artificial air (80% N_2, 20% O_2). The activation energies for AsH_3 adsorption were calculated from the initial rate of change of surface coverage and were found to coincide with hydrogen adsorption. This might indicate that the decomposition of arsine is not the rate-limiting step in the detection process. The sensor further showed sensitivity to NH_3, but no study of the stability of the device was presented. Evans et al. (1986) used a conductance method of E. H. Nicollian and A. Goetzberger (see Nicollian and Brews, 1982) to show that interface state densities vary upon exposure to hydrogen for devices, with thin (~ 12 nm) oxides. This was attributed to interface traps at or very close to the $Si–SiO_2$ boundary, which may be effectively removed by a low-temperature annealing process. No change in the interface state density was observed with thick (~ 100 nm) oxides. More detailed studies of these phenomena were performed both experimentally and theoretically by Fare and Zemel (1987). These authors showed that hydrogen atoms produced by the catalytic action of the Pd on hydrogen molecules can be injected into the oxide–semiconductor interface, where they may modify the density and capture cross sections of the hydrogen-induced interfacial states. Below 125 °C, the injected hydrogen can be removed reversibly by changing the ambient gas from the $H_2 + N_2$ mixture to pure oxygen. Lalauze et al. (1988) have raised the possibility that the Pd–MOS structure stability can be affected by the surface states of thin Pd films, which, when contaminated with impurities such as carbon, sulfur, and oxygen, can affect the active hydrogen adsorption sites and thus may interfere with the diffusion process itself. Petersson et al. (1985) studied under HV and UHV conditions the dependence of the hydrogen heat of adsorption of the internal interface of a MOS structure on the hydrogen coverage. They concluded that there exists a spectrum of hydrogen adsorption sites at the internal surface. This explains the ability of Pd–MOS structures to respond to hydrogen over many orders of magnitude of applied pressure. The clean outer Pd surface was found to have a smaller heat of adsorption under HV conditions than the inner interface and thus is not sensitive to small pressure variations $\leqslant 1 \times 10^{-11}$ torr. When contaminated, however, the outer surface responds fairly similarly to the inner interface. The Pd–MOS sensor's hydrogen energetics is summarized in Fig. 3.50, consistent with a Temkin isotherm, i.e., a coverage-dependent heat of adsorption at the $Pd–SiO_2$ interface. A complete physical analysis of the hydrogen energy manifold in the Pd–MOS matrix may possibly be approached upon consideration of the Pd energy-band structure through its interactions with the hydrogenic potential. Although such an analysis hinges on the understanding of the energy structure of Pd thin layers and on the energy states of the underlying oxide, a quantum mechanical approach similar to that of Lenac et al. (1987) may prove valuable in predicting hydrogenic adsorption state energies in Pd.

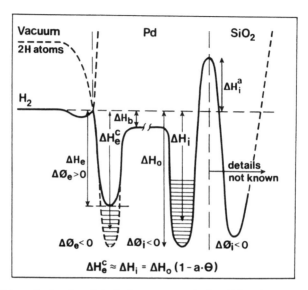

Figure 3.50. Schematic drawing of the hydrogen energetics in a Pd–MOS structure consistent with experimental results: $\Delta H_e \approx 1\,eV/molecule$; $\Delta H_o \approx 1.4\,eV/molecule$, with $a = 1$; ΔH_e^c is the heat of adsorption of hydrogen adsorption sites at the external surface and in the presence of contamination. It is very likely that the activation barrier ΔH_i^a is dependent on the details of sensor fabrication. The hydrogen sticking coefficient on the external surface is close to 1 (Petersson et al., 1984). There may also exist subsurface states (Behm et al., 1983) with heats of adsorption smaller than ΔH_e (Petersson et al., 1985).

3.5.3.2. Device Fabrication and Performance

The history of MOSFET fabrication for gas sensor applications, the chemically sensitive FETs, is largely that of standard microelectronic transistor fabrication, with the exception of the chemically active gate. The first such devices were made as simple n-channel MOS transistors (Lundström et al., 1975a, b). They had very thin oxide and palladium layers ($\sim 10\,nm$ each). Complete sensors were made by mounting these transistors on transistor headers with an externally attached heater and temperature sensor (Stiblert and Svensson, 1975). Although the first available transistors were of bipolar design, the aforementioned chemically sensitive devices were unipolar FETs. Recently, details of some bipolar transistor designs with chemically sensitive properties have been published (Abdullayev et al., 1987; Okada et al., 1987). These were isolated experiments, and their operating mechanisms have not yet been completely understood. In general, Figs. 3.12–3.14 dipict very accurately the fabrication and performance characteristics of mainstream catalytic-gate

MOSFET devices. As was discussed in the case of MOS capacitor sensors, the use of different catalytic metal gates and (depending on the thickness and porosity of the gate electrode) FET-based gas sensors can be made sensitive to various gases besides hydrogen, such as ammonia (Lundström et al., 1986b), or alcohols and unsaturated hydrocarbons (Ackelid et al., 1986).

In what follows particular attention will be paid to a few special types of gas-sensitive MOS devices that do not conform to the conventional device design yet appear to be quite promising for future applications. Jelley and Maclay (1987) fabricated a Pd-gate MOS sensor on p-type silicon wafers with ultrathin Pd layers, 25 and 40 Å thick with an oxide thickness of 100 Å. Contacts were made to allow measurement of the MOS capacitance and impedance across the gate film (Figure 3.51). The device showed sensitivity

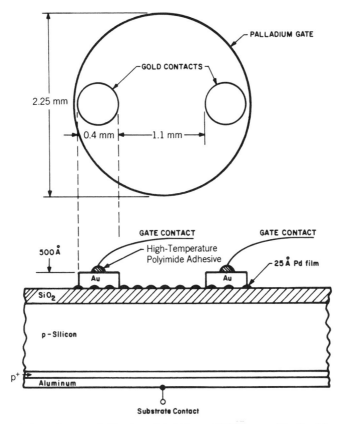

Figure 3.51. Schematic top and side view of the Pd-gate MOS sensor with ultrathin gate metal. Electrical contacts for measuring the gate-film impedance and MOS capacitance are shown (Jelley and Maclay, 1987).

to H_2 at room temperature and to CO and H_2 at elevated temperatures. When measured as a function of time, the changes in MOS capacitance tended to track the changes in the impedance. An effect similar to hydrogen-induced drift (HID) was observed for CO at elevated temperatures. Room-temperature ($T = 24\,^\circ$C) sensitivity to a few ppm of H_2 in air was achieved by the 25 Å Pd-gate device impedance measurement. The threshold voltage shift exhibited similar sensitivity. At 150 °C the threshold voltage shift and impedance were sensitive to ca. 100 ppm CO in air; however, at high CO concentrations (10,000 ppm) a long-term drift of the signal was registered. It is probable that the 25 Å film has an island structure with some coalescence. Quantum mechanical tunneling is thought to be the dominant conduction mechanism in island films, although it has also been suggested that thermionic emission may play a role, particularly at elevated temperatures. The impedance change is thought to be due to the change in the potential between the Pd islands, which affects the tunneling. The 40 Å gate device exhibited lower sensitivities, but also shifted less when exposed to 10,000 ppm CO in air or argon. The 25 Å device also showed sensitivity to 1000 ppm methane in argon, whereas the 40 Å device did not (Maclay et al., 1988). The difference in response for the two film thicknesses is thought to be due to the difference in morphology of the two films: the 25 Å film has an island structure, which allows the CO to alter the potential of the Pd–SiO$_2$ interface; the 40 Å film is more nearly continuous and inhibits the CO from affecting the potential at the interface.

Another interesting FET oxygen sensor using a solid electrolyte was presented by Miyahara et al. (1988). The sensor was fabricated by depositing a thin layer of yttria-stabilized zirconia (YSZ) on a gate insulator of an IGFET (Fig. 3.52). The potential change produced at the interface between the YSZ layer and a Pt-gate electrode (10 nm thick), shows catalytic activity toward oxygen dissociation. A stable signal was obtained, even if the YSZ impedance was very high. The response of the sensor to O$_2$ in N$_2$ was measured at 20 °C by using a specialty circuit that is appropriate for ion-sensitive field effect transistors (ISFETs). A linear relationship between output voltage and the logarithm of $p(O_2)$ was obtained in the range from 0.01 to 1 atm. The sensitivity of the sensor was found to depend on the thickness of the Pt-gate electrode and on the sputtering conditions of the YSZ layer. Although selectivity to hydrogen and carbon monoxide was not good at room temperature, it might improve at higher operating temperatures. The developed sensor has potential advantages over conventional oxygen sensors owing to its small size, low output impedance, solid state construction, and integrability with other semiconductor sensors. It is potentially applicable to medical uses, process control, and automobiles.

Finally, Fig. 3.53 shows the schematic diagram of a new top-gate amor-

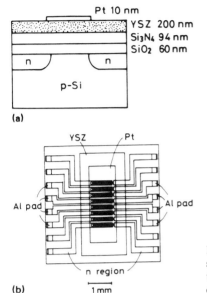

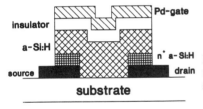

Figure 3.52. Structure of the FET-type oxygen sensor incorporating an yttria-stabilized zirconia (YSZ) thin layer: (a) cross section; (b) layout (Miyahara et al., 1988).

Figure 3.53. Schematic diagram of a Pd-gate a-Si:H thin-film transistor as a hydrogen sensor. The substrate can be glass, kapton, etc. (Mariucci et al., 1990).

phous hydrogenated silicon (a–Si:H) thin-film transistor with a Pd gate and Si_3N_4 as gate insulator. This device has been fabricated and successfully operated as an H_2 sensor (Mariucci et al., 1990). The mechanism of the device has been shown to be similar to that put forth by Lundström et al. (1975a) for the Pd–MOSFET. Responses to 1% and 0.1% $H_2 + N_2$ gas mixtures were measured at room temperature with the transistor operating in the common drain configuration. These responses were similar to that of the Pd–MOSFET (Lundström, 1981; Lundström et al., 1975a). The use of silicon nitride as a gate insulator reduces the diffusion of hydrogen toward the insulator/a-Si:H interface. Therefore, this type of device shows improved long-term stability of H_2 detection. Other advantages include low-temperature processing, relatively simple technology, various kinds of substrates (glass, kapton, etc.), and the possibility of fabricating arrays over a large area.

3.5.3.3.   MOS Device Integration

Recent developments in catalytic-gate MOS device sensor technology have been directed toward the implementation of "intelligent" sensor systems, that is, the fabrication of multiple integrated sensor arrays for pattern recognition in gas sensing. In the mid-1980s Müller and Lange (1986) reported an early temperature-modulated array of MOS gas sensors for gas analysis. At the same time Lundström and Svensson (1985) reported the integration of the Pd-MOSFET into a chip (Fig. 3.54) and its use as a measuring instrument (Fig. 3.47). In 1990 Sundgren et al. reported an "intelligent" gas sensor array with three pairs of Pd-gate MOSFETs and Pt-gate MOSFETs (Fig. 3.55). The array was exposed to a multicomponent gas mixture. Each pair was operated at a different temperature. The signals from the six sensors were recorded by use of calibration and test gas sets that comprised H_2, NH_3, C_2H_4, and C_2H_5OH in 21.3% oxygen in argon as the carrier gas. Signal analysis was performed using both linear and nonlinear PLS (partial least square) models (Zaromb and Stetter, 1984). The predictions/calculations of the PLS models were based on the sensor signals obtained from the calibration gas-set measurements. Figure 3.56 illustrates how the various PLS models predict the hydrogen concentration. The best-fitting model was the modified (m-)PLS model. A smaller scale analysis of a binary gas mixture with two types of MOS gas sensors was successfully carried out by Gall and Müller (1988). Dalby and Shurmer (1988) also presented pattern recognition systems for applications to gas sensor arrays. Such systems, including PLS and the quantitative analysis of gas mixtures carried out with nonselective gas sensors

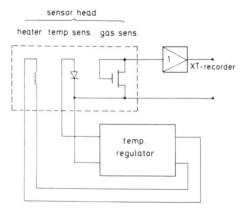

Figure 3.54. Schematic of the integrated circuit in Fig. 3.47 and its connection to the measuring equipment (Lundström and Svensson, 1985).

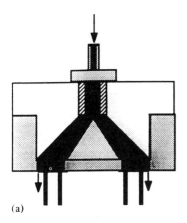

(a)

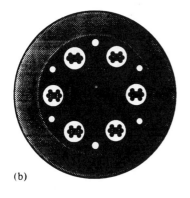

(b)

Figure 3.55. The sample cell with six separate MOSFET sensors, each mounted on a TO-18 socket. The drawing shows the test gas passage and the small dead volume. A cross-sectional view from (a) the side and (b) the top of the sample cell (Sundgren et al., 1990).

(Sundgren et al., 1990; Gall and Müller, 1988; Hierold and Müller, 1988), form the basis of the pattern recognition methods, like PLS, which are expected to become very powerful in the future, especially in the context of industrial applications and environmental studies.

In the meantime, pattern recognition has been extended to the use of the suspended-gate field effect transistors (SGFETs) in identifying two gaseous species (i.e., H_2 and NH_3) independently (Peschke et al., 1990). A SGFET was modified by electrochemical deposition in order to obtain SnO_x and Pd active layers and was tested in a special chamber that could be kept at temperatures between 25 and 200 °C. Gases were diluted by mixing with flow-rate controllers or filling a vacuum chamber. The SnO_x-modified SGFET exhibited a logarithmic response to NH_3 from 50 to 5000 ppm. No sensitivity was observed upon exposure to H_2, CO, or CH_4. The operating temperature range was 35–100 °C. On the other hand, the Pd-modified SGFET showed a reproducible response to H_2 and NH_3. When synthetic air was used as a

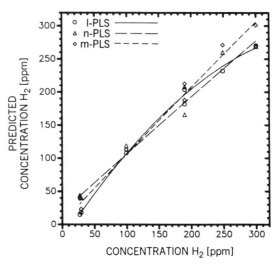

Figure 3.56. Predicted vs. actual hydrogen concentration in the test set for the linear (l-)PLS, nonlineal (n-)PLS, and modified (m-)PLS model (Sundgren et al., 1990).

carrier gas, a combination of both types of SGFETs allowed H_2 and NH_3 to be detected independently. This type of sensor system has a strong potential to be used as a practical multisensor.

Sensor integration has been further utilized not only as a pattern recognition system but also in efforts to improve the response signal to a single gaseous component. This was recently achieved by integrating into a single chip several identical gas-sensitive MOSFETs of the conventional type (Lundström, 1981) into a series-connected circuit (Dobos et al., 1990). Under this configuration, the threshold voltage of the series circuit of n MOSFETs is the sum of each individual element:

$$V_T = \sum_{j=1}^{n} V_T^{(j)} \tag{3.117}$$

Therefore, the potential shifts due to the presence of an active gas concentration are also summed:

$$\Delta V_T = \sum_{j=1}^{n} \Delta V_T^{(j)}, \tag{3.118}$$

which considerably improves the sensitivity of the device. Figure 3.57(a) shows transistors connected in series, and Fig. 3.57(b) is a top view of a chip

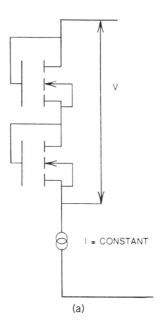

(a)

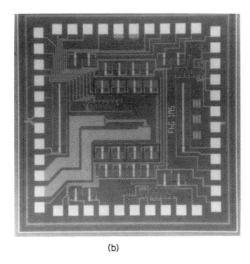

(b)

Figure 3.57. (a) Series connection of two MOSFET gas sensors. (b) Integrated MOSFET sensor chip (Dobos et al., 1990).

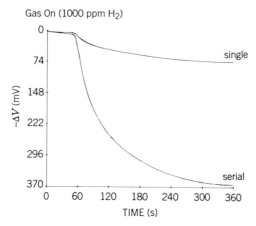

Gas On (1000 ppm H$_2$)

Figure 3.58. Gas concentration measurement with one and four serially connected Pd–MOSFET gas sensors (Dobos et al., 1990).

with two large gas sensors, twenty small, connected gas sensors, three linear temperature sensors, and heating elements located on different areas of the chip. The improved sensitivity manifests itself in the sense of dynamic range increases as shown in Fig. 3.58. A negative by-product of MOSFET sensor integration was the exhibited signal drifts and sensitivity to the background ambient conditions (e.g., humidity), stronger than a single sensor. Nevertheless, the serial connection generally improved the signal-to-drift ratio.

3.5.4. MiS Diode Sensors

3.5.4.1. General Considerations

An extensive early study of the Pd–thin SiO$_2$–Si structure was that by Keramati (1980). An excellent introduction to this type of MiS diode structures was written by Zemel et al. (1981). Fabrication of MiS structures grew out of necessity so as to reverse the failure of early Pd–Si diodes to detect hydrogen. This was attributed to a Pd$_2$Si layer forming during the fabrication of Pd–Si Schottky diodes (Kricher, 1971; Bower and Mayer, 1972). A thin grown oxide layer was shown to provide an effective barrier against the formation of the silicide intermediate layer, thus allowing the diode to be sensitive to H$_2$. The small admittance of the signal from the highly sensitive Pd–thin SiO$_2$–Si diodes was a clear indication that the operating mechanism of H$_2$ sensitivity cannot be solely due to a change in the metal work function (Steele et al., 1976) or the appearance of a dipole layer (Shivaraman et al., 1976) at the Pd–SiO$_2$ interface. According to Zemel et al. (1981), the effect

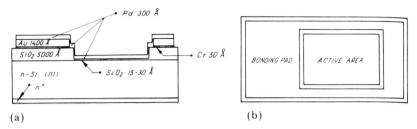

(a)   (b)

Figure 3.59. Schematic diagram of a Pd-thin SiO_2–Si diode. The 0.93 mm^2 active area consists of a $\langle 111 \rangle$ 2–4 $\Omega \cdot$cm silicon substrate, a 15–30 Å oxide and a 300 Å Pd layer. The oxide was grown in air at 200 °C for $\frac{1}{2}$ h. The bonding pad is isolated from the substrate by a 5000 Å thick oxide layer. A 1400 Å gold layer cushions the Pd and allows good contact with the Pd electrode. (a) Cross-sectional view. (b) Top view (Keramati, 1980).

of hydrogen on the small signal admittance response of the diode had to be associated with the appearance and removal of surface states at the Si–SiO_2 interface, as well as a possible work function shift at the Pd–SiO_2 interface. A schematic diagram of a Pd-thin SiO_2–Si diode is shown in Fig. 3.59. Figure 3.60 illustrates the current–voltage (I–V) behavior of such a diode under pure oxygen and 100 ppm hydrogen diluted in nitrogen. The time response of the

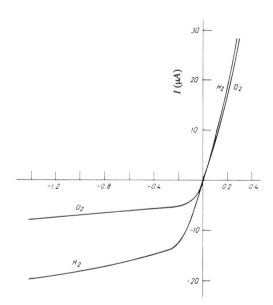

Figure 3.60. Current–voltage characteristics of a Pd-thin SiO_2–Si diode under pure oxygen (O_2) and 100 ppm hydrogen diluted in nitrogen (H_2) at 108 °C (Zemel et al., 1981).

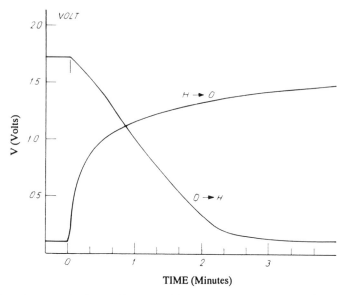

Figure 3.61. Oxygen-to-hydrogen (O→H), and hydrogen-to-oxygen (H→O) time response of the reverse bias voltage of the Pd-thin SiO$_2$-Si diode of Fig. 3.60, under conditions of pure O$_2$ (O→H), and 100 ppm H$_2$ in N$_2$ (H→O) at 108 °C (Zemel et al., 1981).

same device is shown in Fig. 3.61, where the reverse bias voltage at constant current is plotted vs. time as the ambient gas is switched from pure O$_2$ to diluted H$_2$ and vice versa. The presence of oxygen has been found to be essential in the removal of hydrogen from the diode structure. For hydrogen concentrations in excess of 1 ppm the full hydrogen response in the absence of oxygen is independent of the p(H$_2$). However, the response of the diode to hydrogen is faster at higher partial pressures and temperatures.

The $(J-V)$ behavior of MiS devices can be described analytically, assuming a conduction mechanism due only to majority carriers, by a modification of Eqs. (3.9), (3.10), and (3.12) (which are strictly valid for true Schottky barriers) in order to account for the presence of the thin oxide layer through a combination of diffusion mechanisms (Wronsky et al., 1976) and oxide tunneling (Card and Rhoderick, 1971):

$$J(V) = \tau A^* \exp(-q\phi_B/kT)[\exp(qV/nkT) - 1] \qquad (3.119)$$

Here τ (< 1) gives the probability of transmission across the oxide layer; A^* is the effective Richardson constant for free electrons defined in Eq. (3.12); and n is the ideality (also known as "quality") factor. Figure 3.62 shows

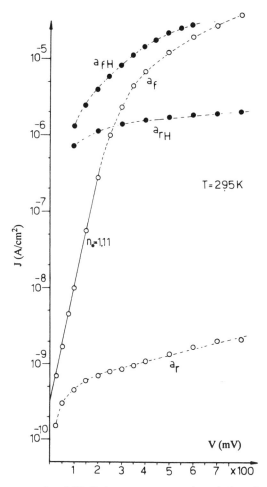

Figure 3.62. J–V curves of an MiS diode structure exposed to air (a_f, a_r) and to 280 ppm of H_2 in nitrogen (a_{fH}, a_{rH}) (D'Amico et al., 1983a).

typical $(J$–$V)$ curves of a Pd–MiS device with $\sim 20 \text{ Å}$ of SiO_2, when exposed to air (a_f, a_r) and to 280 ppm of H_2 in a hydrogen-nitrogen mixture (a_{fH}, a_{rH}). From the slope of the linear region of a_f, as plotted in Fig. 3.62, and from its extrapolation to zero bias, n and $J_0 \equiv \tau A^* \exp(-q\phi_B/kT)$ values have been evaluated. For the air-exposure, room-temperature data of Fig. 3.62, $n = 1.11$ and $J_0 = 3 \times 10^{-10} \text{ A/cm}^2$ (D'Amico et al., 1983a). The diodes fabricated by Keramati (1980) exhibited ideality factors $n = 4.89$ under pure O_2, and $n = 5.89$ under 10% H_2 in N_2 (Zemel et al., 1981). Such unusually high values clearly indicate that a simplistic model is unacceptable and the barrier heights obtained through the Schottky diode analysis are not to be trusted. This conclusion and other electrical testing results on small-signal equivalent capacitance and conductance are consistent with the earlier remark that the hydrogen effect cannot be merely associated with the metal work function shift. To address the important data interpretation problem, D'Amico et al. (1983b) performed studies of the electronic transport in MiS diodes in the dark, in air, or in the presence of $N_2 + H_2$ fluxes with various concentrations of H_2 near atmospheric pressure conditions. The substrate semiconductor was amorphous hydrogenated silicon (a–Si:H) in this case. The barrier height values were determined by the temperature dependence of the saturation current, while the built-in potential was evaluated from the $(C–V)$ relationship. These authors used Eq. (3.118) for the interpretation of their data with (Wronsky et al., 1976).

$$A^* = q\mu_c N_c E_s \qquad (3.120)$$

where μ_c is the free electron mobility in the conduction band; N_c, the effective density of states; and E_s, the electric field at the surface of the semiconductor. The mechanism for the Pd–thin SiO_2–a-Si:H diode sensor sensitivity to H_2 was consistent with a change in the contact potential V_c, which was reflected in a reduction of both the barrier height, ϕ_B, and the built-in potential, V_B. This type of behavior also agrees with the fact that in a-Si:H the density of states in the gap is so high that any significant further increase due to H_2-induced trapping states seems improbable. The measurements of ϕ_B and V_B further allowed the evaluation of the position of the Fermi level in the a-Si:H semiconductor material used. This was located at ca. 0.5 eV below the conduction band, in agreement with the activation energy value $\Delta E \simeq$ 0.53 eV, obtained by the series conductance analysis of the diode. A most effective technique for the investigation of trapping centers in the MiS structure has been proven to be deep-level transient spectroscopy (DLTS) (Lang, 1974). When this technique was applied to Pd–thin SiO_2–(n-type Si) diodes exposed to H_2, two electronic traps were revealed, both near the Si surface (Petty, 1982). It was subsequently suggested that one of these traps may be related to the hydrogen dipole layer directed toward the SiO_2–metal interface (Lundström, 1981). Unfortunately, no other DLTS applications to solid state gas sensors appear to have occurred to date. In our opinion a wealth of information on the energetics and origin of the diode signal can be garnered from this technique, contributing greatly toward sensor mechanism elucidation. The only potential drawback of DLTS is the relatively sophisticated instrumentation and signal-processing requirements.

Fortunato et al. (1984) investigated the Pd–SiO_x–a-Si:H hydrogen sensor further by photoemission spectroscopy with synchrotron radiation. In order to overcome difficulties in applying this technique at the metal–oxide interface owing to the dominant metal d-electron emission even for ultrathin Pd layers, these workers irradiated the a-Si:H region of the structure instead. From the photoemission experiments the occupied electronic state densities as functions of energy relative to the Fermi level were estimated, and this information was then used to construct the energy-band diagram of Fig. 3.63: In the as-grown system [Fig. 3.63(a)], the top of the valence band of the amorphous silicon coincides with the position of E_F and $\Delta E_V = 2.5\,eV$. The Fermi level has been placed in the middle of the gap of the thin oxide layer, which is so thin (10–15 Å) that it can be practically represented by flat bands. After hydrogenation the new equilibrium situation is shown in Fig. 3.63(b). The top of the SiO_x valence band is shifted by ΔE, and $\Delta E_V = 2.1\,eV$ is reduced. The a-Si:H has a larger gap than unhydrogenated a-Si and is assumed to be intrinsic. When the hydrogen is removed (upon exposure to O_2), the top of the SiO_x valence band reaches its original position, while ΔE_V remains constant. This results in a net increase of the Schottky barrier height, ϕ_B [Fig. 3.63(c)). The photoemission study thus concluded that the main mechanism responsible for the variation in the transport properties of Pd–thin SiO_x–(a-Si:H) structures under hydrogen exposure is a change in

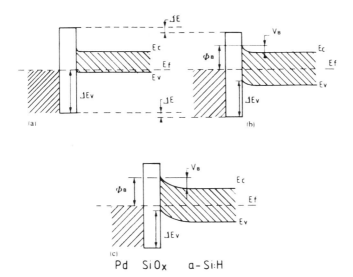

Pd   SiOx   a- Si:H

Figure 3.63. Band scheme of the Pd–SiO_x/a-Si system relative to the density of occupied states spectra obtained from photoemission spectroscopy. A thick a-Si layer is shown (a) after a-Si deposition, (b) after hydrogen exposure, and (c) after oxygen exposure. The depletion region width is arbitrary (Fortunato et al., 1984).

the contact potential, in agreement with the earlier conclusions (D'Amico et al., 1983b).

Perhaps the most thorough mechanistic study to date with respect to the operation of the hydrogen sensing in MiS diodes has been presented by Butler (1985). Pd–thin oxide–Si structures with $t_{SiO_2} \simeq 22$ Å were fabricated and were studied using capacitance, conductance, and photo-induced current measurements. The data obtained by Butler was indicative of two processes: (a) majority carrier tunneling from the Pd through the oxide to the Si–SiO$_2$ interface states (which were hypothesized to be present at significant densities); and (b) thermionic emission over the Schottky barrier. This behavior can be modeled by a metal–semiconductor diode in series with a resistor. When applied to this structure, Eq. (3.119) can be written as

$$I = A^* A T^2 \exp(-q\phi_B/kT)\exp(-\alpha\chi^{1/2}d)[\exp(qV/kT)-1] \quad (3.121)$$

with A being the diode area; χ, the oxide barrier height (in eV); d, the oxide thickness; and $\alpha \simeq 1$, an electron tunneling constant. When differentiated, the Card and Rhoderick (1971) expression, Eq. (3.118), gives the diode resistance at zero bias. The resistance obtained from data extrapolation to zero barrier height ($\phi_B = 0$) is due to the tunneling resistance of the oxide layer. The oxide-tunneling and thermionic emission model can be described by the composite resistance

$$R = R_0 \exp(\chi^{1/2}d) + \left(\frac{k}{qA^*AT}\right)\exp(q\phi_B/kT) \quad (3.122)$$

where R_0 was found to be equal to 700 through data fitting to the diode resistance as a function of Schottky barrier height at zero bias. It was the exposure to $p(H_2)$ (0–100 ppm in N$_2$) that modified the barrier height. These results strongly support the importance of interface states at the Si–SiO$_2$ interface in the conduction process of MiS diodes. An upper limit on the interfacial density of states at the Si–SiO$_2$ interface was estimated to be 10^{12} states/cm^2·eV, and the Schottky diode–resistor model was found to be capable of describing the equilibrium characteristics of the Pd–MiS diode when exposed to various concentrations of H$_2$.

Photo-induced current measurements showed both photoionization of deep levels in the depletion layer and photoemission from the Si–SiO$_2$ interface states. The photoemission confirmed the presence of these interface states, and the value of the Schottky barrier height was measured capacitively. An expression for the spectral dependence of the photoemission was derived (Butler, 1985) based on the photo-assisted tunneling concept of Burshtein and

Levinson (1975):

$$J = C \int_{E_F + q\phi_B}^{E_F + hv} D(E_x)(E_F + hv - E_x)dE_x \qquad (3.123)$$

where C is proportional to the intensity of light; E_x, the x-component of the kinetic energy of the electrons; and $D(E_x)$ the tunneling probability of those electrons. Upon writing this quantity as

$$D(E_x) = \exp[-B(E_f + \chi - E_x)^{1/2}] \qquad (3.124)$$

with

$$B = (4\pi d/h)(2m_e^*)^{1/2} \qquad (3.125)$$

Butler obtained for the photo-induced current density:

$$J(v) = \frac{6C}{B^2}(hv - q\phi_B)\exp[-B(\chi - q\phi_B)^{1/2}] \qquad (3.126)$$

where the constants in B include the oxide thickness, d, and the electron effective mass, m_e^*. The spectral dependence of J shows the effect of the oxide on the photoemission process and the reason for which such a process is not observable with thick-oxide MIS sensors. Butler thus concluded that changes in current under reverse bias upon exposure to H_2 result from lowering of both the Schottky barrier and the oxide barrier height, but not from changes in the Si–SiO$_2$ interface state density.

On the other hand, there is growing evidence that understanding the physics of interfacial states in the oxide is extremely important for interpreting many facets of the MiS sensor behavior. Keramati and Zemel (1982a,b) pointed out that Pd gates can act as reversible "contacts" for isothermal injection of atomic hydrogen, H, into the Si–SiO$_2$ interface of a Pd–MOS capacitor sensor. De Rooij and Bergveld (1978) further investigated the influence of hydrogen on MOS sensors by examining directly the effect of pH on the interface state density. They found that the density decreased with decreasing pH but could not separate out the effects of hydrogenic species from those of other chemical species in solution. Of direct relevance to MiS structures, Wen and Zemel (1979) demonstrated that when 100 nm thick silicon oxides of an ion-controlled diode are exposed to acids and bases, a substantial increase in mid gap surface recombination and interfacial trapping states occurs. Furthermore, they found that the surface recombination velocity had increased dramatically as a consequence of an increase in the number of interfacial trapping states located in the same general E_c–0.4 eV region as the hydrogen-induced states observed by Keramati and Zemel (1982a, b). These latter authors employed a Pd–thin SiO_2–Si diode to study the isothermal reversible injection of H into the Si–SiO_2 interface. They found (a) an increase in tunneling current associated with the expected Pd work function, ϕ_{m-s}, change, and (b) changes in the tunneling current and admittance corresponding to changes in the interface state properties. However, their oxide was an ill-characterized native oxide arising from a short time exposure of a freshly etched Si wafer to the laboratory air at 100–120 °C. Other investigators failed to observe these changes in the interfacial state densities or capture cross sections on Pd–MiS diodes (e.g., Ruths et al., 1981). Fare and Zemel (1987) attribute these apparent inconsistencies partly to the fact that different investigators employed different measurements and therefore no direct correlation of results could be made. The observations by Keramati and Zemel (1982a,b) were, nevertheless, corroborated by Fare et al. (1986), who not only showed that hydrogen can indeed be injected into SiO_2 but also that it can cross the far SiO_2–Si interface and deactivate bulk acceptor (boron) states. Much progress has been achieved recently in a thoroughly careful experimental and theoretical analysis, in which Fare and Zemel (1987) established that hydrogen can be reversibly injected into a (very-well-characterized) Si–SiO_2 interface by using a Pd metal gate. This hydrogen can perturb the interfacial state density profile, producing an excess of states in the vicinity of the E_c–0.4 eV region (Keramati and Zemel, 1982a, b). In oxides prepared using various methods [i.e. trichloroethane (TCA), doped; dry oxide; and wet oxide], the data obtained from conductance changes demonstrated that hydrogen can diffuse through as much as 10 nm of SiO_2 at low-to-moderate temperatures, in agreement with Fare et al. (1986). The hydrogen injection/extraction experiments were fully reversible for $T < 125$ °C and can be combined with standard interfacial state measurements to characterize Si–SiO_2 interfaces prepared in different oxidizing atmospheres. The reversible injection of hydrogen into the Si–SiO_2 interface has been tentatively interpreted in terms of the strain at the interface produced by the additional hydrogen, proposed theoretically by Singh and Madhukar (1981), while additional calculations taking into account the details of the Si–H–O interaction appear to be necessary in order to obtain more quantitative information for comparison with experiments.

The results obtained by Keramati and Zemel (1982a) indicate that hydrogen penetrates the oxide layer and participates in the formation of the hydrogen interfacial trapping states in ways that may depend on the method of oxide growth. This finding can raise questions regarding the long-term stability and reliability of MiS diodes as hydrogen sensors. In the short term, hydrogen effects can be reversed by oxygen systematically and no degradation of the device performance has been observed. Formation of interfacial traps appears to be a mechanism for the sensitivity to hydrogen of the Pd–MiS diode, which has exhibited considerable potential as a practical device although the question of stability requires further examination. In terms of sensitivity, Pd–MiS diodes with ultrathin (~ 30 Å) thermally grown SiO_x layers between the Pd and Si have been known to exhibit enhanced response to the presence of hydrogen (Ruths et al., 1981), provided that the junction is carefully formed using controlled oxidation of the silicon and thermal evaporation of the palladium. The $J–V$ characteristics of such a diode, before and after exposure to 154 ppm of hydrogen in nitrogen, are shown in Fig. 3.64 and should be compared to the $J–V$ curves of a conventional MiS diode (Fig. 3.62). The characteristics in Fig. 3.64 show that much larger currents are passed by a Pd/SiO_x/(n)Si MiS diode, at a given voltage, when the diode is exposed to H_2. It is believed (Fonash et al., 1982) that the enhanced sensitivity is because the thermionic emission component of Eq. (3.118) dominates the $J–V$ characteristics of very sensitive devices, such as the one described in Fig. 3.64, and because the presence of hydrogen can modify ϕ_B. This response is essentially

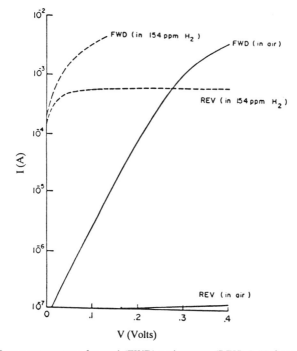

Figure 3.64. Room-temperature forward (FWD) and reverse (REV) $J–V$ characteristics of an MiS Pd–SiO_x/(n)Si diode in air and in 154 ppm H_2 in N_2 (Fonash et al., 1982).

reversible. Possible mechanisms for ϕ_B modification involve hydrogen dissociation and absorption lowering the work function of Pd (Lewis, 1967): the atomic hydrogen produced by the Pd acting on H_2 gives rise to a dipole layer between the metal and the semiconductor (or ultrathin insulator), which effectively lowers the metal work function; or absorbed hydrogen reacts with Pd to form a hydride with lower ϕ_m (Steele et al., 1976), thus lowering ϕ_B, ideally through (Rhoderick, 1978)

$$\phi_B = \phi_m - \chi \qquad (3.127)$$

or the hydrogen modifies intrinsic states or extrinsic states, or both, present in the interface, thereby modifying the barrier and consequently ϕ_B. This process may also be explained through the modification of atomic dipoles in the interface by hydrogen, which would result in modifying the barrier, and ϕ_B (Yamamoto et al., 1980). This type of mechanism differs drastically from the aforementioned one proposed by Keramati and Zemel (1982a), who suggested that more current is passed in the presence of hydrogen, because the current density J_{IS}, due to electron recombination or capture at the interface, increases as a result of the increased interface state density. That explanation was consistent with the behavior of diodes that were much less sensitive to H_2 than those exemplified in Fig. 3.64. In the $Pd/SiO_x/(n)Si$ diodes of Fig. 3.64 the opposite was observed (Fonash et al., 1982): in the presence of hydrogen, devices made on p-type Si became more rectifying. This is consistent with the explanation that hydrogen lowers the barrier height on n-type Si and therefore raises the barrier height on p-type Si substrate. Rye and Ricco (1987) addressed the problem of Pd metal surface contamination in Pd–MiS H_2 sensors by performing both steady state and kinetic experiments with diodes with clean Pd surfaces under vacuum conditions and hydrogen pressures ranging from 10^{-10} to 10^{-1} torr. In such a clean environment, the lower limit $p(H_2)$ to which MiS diodes with a 20 Å thick SiO_2 layer exhibited a response was at least 7 orders of magnitude more sensitive than that obtained for devices with contaminated surfaces. The presence of controllable vacuum conditions, down to the sensitivity limit of 10^{11} total H_2 impacts/cm^2 at the Pd surface, allowed these workers to probe the H_2/Pd interaction mechanism more effectively than the aforementioned experimental work performed under STP conditions. Important conclusions from the work by Rye an Ricco (1987) are as follows:

1. Transport of hydrogen to the interface (or to the bulk) cannot occur exclusively through a strong chemisorbed surface species. The general characteristics of the kinetic data were found to be consistent with a mechanism involving a weakly bound precursor state, summarized in the chemical

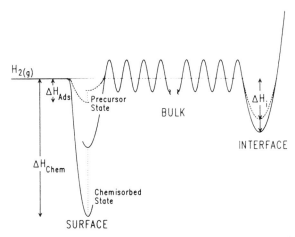

Figure 3.65. Potential energy diagram for the precursor model given in Eqs. (3.128). Indicated are the enthalpies of physisorption on Pd, ΔH_{ads}; chemisorption on Pd, ΔH_{chem}; and enthalpy of binding at the Pd/SiO$_2$ interface, ΔH_i. Energies are to scale (Rye and Ricco, 1987).

reaction scheme:

$$H_{2(g)} \xrightleftharpoons{1} 2H_{(a)} \begin{array}{c} \xrightarrow{2} 2H_{(b)} \xrightleftharpoons{3} 2H_{(i)} \\ \updownarrow 5 \\ \xleftarrow{4} 2H_{(c)} \end{array} \qquad (3.128)$$

The subscript a in Eqs. (3.128) refers to the weakly absorbed precursor. The subscripts g, c, b, and i denote gas-phase, chemisorbed, Pd bulk, and Pd–SiO$_2$ interfacial hydrogen, respectively. The qualitative features of the precursor model are summarized in the activation energy diagram of Fig. 3.65.

2. Adsorption of hydrogen at the surface could be rate limiting over the lower portion ($p < 10^{-7}$ torr) of the H$_2$ pressure range, but only as a result of a low impact rate. At pressures where the rates for the individual steps are comparable, chemisorption of hydrogen may indirectly affect the rate by an interdependence of the precursor and chemisorbed hydrogen concentrations.

3. The steady state diode signal is consistent with two interface states, the binding enthalpies of which are 9 and 6.8 kcal/mol of H with respect to H$_{2(g)}$ (Fig. 3.66).

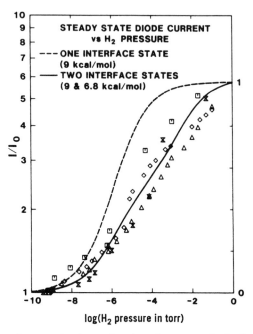

Figure 3.66. Near steady state signal as a function of pressure obtained at the end of each pressure jump from the reverse-bias current response vs. time at $p(H_2)$ covering the pressure range 6×10^{-10} to 8×10^{-2} torr. Points represent data from two different diodes. The left vertical scale refers to the experimental points, and the right vertical (linear) scale refers to the calculated interface $[H_{(i)}]$ coverage (Rye and Ricco, 1987).

The discussions presented here clearly show that the physics and chemistry of gas detection in MiS diodes is far from well understood and that much more work must be carried out for a unified theory to emerge, especially under UHV conditions, as demonstrated by the very promising results by Rye and Ricco (1987).

3.5.4.2. MiS Device Integration

The state of the art in MiS-based gas sensor device integration rests with the developments reported by Tatsuo Yamamoto's group at Shizuoka University, Hamamatsu, Japan. A hydrogen-switching sensor chip was reported by Ogita et al. (1986). This sensor was based on a MiS tunnel structure that exhibits switching characteristics (Yamamoto and Morimoto, 1972), as shown in Fig. 3.67. A $p-n$ junction is in series with a thin insulating layer of 2–4 nm SiO_2, through which tunneling-induced currents can flow. The $C-V$ characteristics

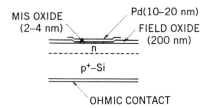

Figure 3.67. Cross section of the hydrogen-sensitive switching Pd–Si tunnel MiS sensor (Ogita et al., 1986).

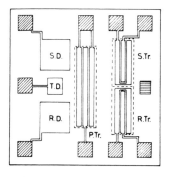

Figure 3.68. Pattern of an integrated hydrogen switching sensor fixed with a heater P.Tr., Power transistor; S.Tr., sensor transistor; R.Tr., reference transistor; T.D., temperature diode; S.D., sensor diode; R.D., reference diode (Ogita et al., 1986).

of the device were those of a current-controlled negative resistance with the effective functionality of a high-speed switch (Kroger and Wegener, 1973). The sensor was found to have a strong dependence on temperature and was therefore deemed necessary to keep the temperature constant in order to obtain high sensitivity and reliability, along with a short response time. An integrated chip has been fabricated (Fig. 3.68) that includes a power transistor as a heater and a diode as a temperature sensor. This device is optimally operated at $100\,°C$ and can detect $10\,ppm$ H_2 in air: the p–n–p power transistor inside the n-on-p^+ Si chip can heat the entire chip up to $100\,°C$ with approximately $500\,mW$ of input power. The $[H_2]$ is detected through a corresponding change in the switching voltage of the device, as it transitions from the very-high-impedance OFF state to a low-impedance ON state, following the tunneling current flow when the p–n junction is forward-biased in the presence of ambient hydrogen. The same group (Murakami et al., 1988) have reported the fabrication of a different integrated chip, which consists of a Pd–Si MiS tunnel diode (hydrogen sensor), a diffused resistor layer (inside heater), and a p–n junction (temperature control). The chip with a device hole-structure Pd layer of $20\,\mu m$ in diameter exhibited a drastic improvement in hydrogen sensitivity, as compared to a device without a hole-structure Pd layer.

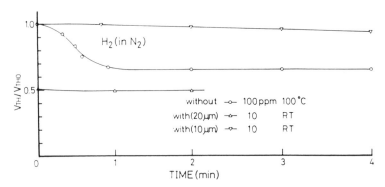

Figure 3.69. Comparison of the response characteristics for the MiS tunnel diodes, with and without a hole structure Pd layer. All voltage curves in the presence of hydrogen, V_{TH}, have been normalized to the switching voltage, V_{THO}, in the absence of hydrogen (Murakami et al., 1988).

For comparison, the responses of the chips without and with hole structures are shown in Fig. 3.69. However, a chip with a device hole structure $10\,\mu m$ in diameter did not show any improvement in $10\,ppm\,H_2$. The results with the $20\,\mu m$ holes in the Pd layer suggest the possibility of room-temperature operation of the detector. Clearly, accelerated rates of development are needed toward the practical implementation of Pd–MiS integrated sensors under ambient conditions.

3.5.5. Metal–Semiconductor (Schottky Diode) Sensors

This family of gas sensors is sometimes included with the larger set of MiS sensors, especially where ultrathin insulating layers are involved. The transition from an MiS structure to an M–S diode structure has been studied on the metal(Ag or Pd)–ultrathin SiO_x insulator–Si(100) system by Sullivan et al. (1978). These authors examined the effects of controlled surface oxidation and subsequent metallization on the characteristics of Schottky barrier-type devices. Using dark current–voltage $(J–V)$, reverse capacitance–voltage $(C–V)$, and photovoltaic characteristics, they observed that the change in V_{OC} in going from the MiS to the M–S structure is caused by barrier height changes in a smooth manner. Once the insulating layer between a catalytic metal film (almost exclusively Pd) and a semiconducting substrate has been removed, the M–S Schottky diode has been found to have gas-sensing abilities provided that the following requirements are met: (a) no undesirable formation of intermediate species through reaction with the metal; and (b) no Fermi-level pinning (Saaman and Bergveld, 1985).

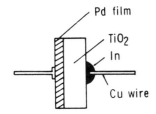

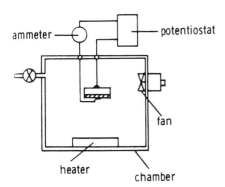

Figure 3.70. A schematic diagram of the Pd–TiO₂ Schottky diode and the measuring cell (Yamamoto et al., 1980).

Hydrogen sensors based on Pd-ZnO (Ito, 1979), Pd–CdS (Steele and MacIver, 1976), and Pd–TiO₂ (Yamamoto et al., 1980) Schottky diodes have been reported. The predominant mechanism responsible for the sensitivity to hydrogen of these M–S sensors has been described as the work function change of the catalytic metal upon chemical interaction with hydrogen gas. The resulting current–voltage curve modification of the Schottky diode has been used to quantify the sensor characteristics. Figure 3.70 is a schematic diagram of the Pd–TiO₂ diode and the measuring cell. We choose this parti-cular device (Yamamoto et al., 1980) as an example of the simplicity of fabri-cation of such a diode. Single crystalline TiO_2, a (001)-oriented wafer, was heated to $500\,°C$ at 4×10^{-4} torr for 1 h and indium was evaporated onto one of the (001) surfaces in vacuum so as to form ohmic contact. Then a copper wire was soldered on. The other (001) surface of the TiO_2 wafer was polished, etched, washed, and dried and Pd was evaporated at 2×10^{-5} torr. The Pd thickness (ca. 200 Å) was monitored with a quartz crystal oscillator. Finally a copper wire was directly pressed on the Pd film. The I–V curves for the Pd–TiO₂ diode in air and in hydrogen-containing air are shown in Fig. 3.71. The effect was explained by the change of the Pd work function, under the action of gases, notably H_2 and O_2. This was further confirmed by the direct measurement of the change of surface potentials of Pd film and

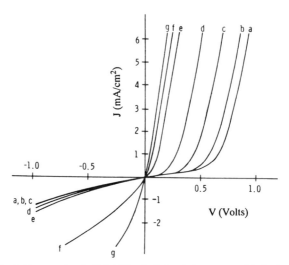

Figure 3.71. The $I-V$ curves for the Pd–TiO$_2$ diode in air (curve a) and in hydrogen-containing air (curves b–g). The partial pressures of hydrogen are 14 ppm (b), 140 ppm (c), 1400 ppm (d), 7150 ppm (e), 1% (f), and 1.5% (g). The temperature of the diode was 25 °C (Yamamoto et al., 1980).

TiO$_2$ crystal separately (leading to the work function change), by use of a vibrating capacitor method at various hydrogen partial pressures. Yamamoto et al. (1980) found that the work function of Pd changed reversibly (decreased) with $p(\mathrm{H}_2)$ whereas that of TiO$_2$ crystal hardly changed. The work function of Pd seems to be controlled by the amount of adsorbed oxygen determined by the rates of the reaction with hydrogen and the trapping of free oxygen from the air (Fig. 3.72). Similar electrical properties were further studied with respect to TiO$_2$ junctions with Pt, Au, Ni, Al, Cu, Mg, and Zn; and of the semiconductors ZnO, CdS, GaP, and Si with Pd (Yamamoto et al., 1980). The Pt–TiO$_2$ and Au–TiO$_2$ diodes had electrical properties similar to those of the Pd–TiO$_2$ diode, but the sensitivity to hydrogen decreased in the order: Pd > Pt > Au. The Ni–TiO$_2$ diode exhibited no sensitivity to H$_2$, nor to other reducing gases (CO, C$_2$H$_5$OH, and C$_3$H$_6$), in the range 25–200 °C. The Pd–TiO$_2$ diode becomes sensitive to these gases at $T > 100$ °C. The junctions of TiO$_2$ with Al, Cu, Mg, and Zn showed ohmic behavior and no rectification even in air. The response of a Pd–ZnO diode in hydrogen-containing air was similar to that of the Pd–TiO$_2$ device. Finally, the responses of CdS, GaP, and Si with Pd were unsuitably small. The probable reason for this poor performance was hypothesized to be the large deviation of these junctions from the ideal Schottky behavior, owing to the chemical reactions

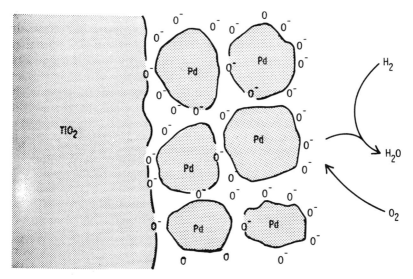

Figure 3.72. A schematic model for the Pd–TiO$_2$ interface (Yamamoto et al., 1980).

of the semiconductor with the metal at the interface (Brillson, 1977, 1978). In the case of Pd–CdS and unlike the results by Yamamoto et al. (1980), Steele and MacIver (1976) did observe a marked decrease of the Schottky diode barrier height by ca. 0.53 eV, which was believed to be due to the Pd work function decrease. The 10^6 $\Omega \cdot$cm single-crystal CdS used was evaporated with Pd dots 800 Å thick and 1 mm diameter and, upon exposure to hydrogen, was found to make a good H$_2$ sensor down to 298 K, over the range 500–5000 ppm H$_2$. Ito and Kojima (1982) used a Pd–very thin SiO$_2$–n-Si Schottky diode (no thickness of the SiO$_2$ was given) operating at room temperature to detect H$_2$ in air. Their sensor was able to respond to 2000 ppm H$_2$ within 10 s. These workers operated their diode sensor in the forward-bias mode and verified that the current could be given approximately by Eq. (3.16) with $n \simeq 1$.

Very recently Nie and Nannichi (1991) reported a H$_2$ gas sensor consisting of a Pd-on-GaAs Schottky contact. Table 3.7 shows typical data on the variation of the barrier height of a chemically etched Pd-on-GaAs contact, the only type of sample which responded to hydrogen. The function ϕ_b was derived from Eq. 3.10, with A^{**} the effective Richardson constant of 8.16 and 74.4 A/cm$^2 \cdot$K^2 for n-type and p-type GaAs, respectively. The Schottky barrier heights were found to depend on the interface structure due to an intermediate layer resulting from the reaction of Pd with GaAs: the barrier was sensitive to hydrogen when its height was greater than a critical value

Table 3.7. Variation of the Barrier Height of a Typical Chemically Etched Pd-on-GaAs Schottky Contact (Before Heating) Due to Hydrogenation at Room Temperature

	Air	H_2 (0.5 atm)
n-Type GaAs		
ϕ_b^{I-V} (eV)	0.95	0.79
ϕ_b^{C-V} (eV)	1.05	0.82
n value[a]	1.09	1.03
p-Type GaAs		
ϕ_b^{I-V} (eV)	—[b]	0.50
ϕ_b^{C-V} (eV)	—[b]	0.70
n value[a]	—[b]	1.20

Source: Nie and Nannichi (1991).
[a]Ideality factor.
[b]Data not available owing to low barrier heights.

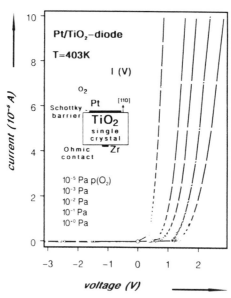

Figure 3.73. Current–voltage curves of Pt–TiO$_2$ Schottky diodes with zirconium as ohmic backside contact after exposure to various partial pressures of oxygen at constant temperature $T = 403$ K (Kirner et al., 1990).

of ca. 0.96–0.99 eV. Unlike other hydrogen sensors, this particular device may be singled out in that it may use hydrogen as a probe of the sensor's interface structure, i.e., whether metallic Pd or an intermediate material consisting of Pd–Ga–As is in contact with GaAs. Aspnes and Heller (1983) had also reported earlier on some sensitivity of GaAs contacts of Ru, Rh, Ir, Os, Pd, and Pt; the latter two metals had failed to respond, a fact attributed to a chemically altered interface structure. Kirner et al. (1990) used TiO_2 thin films rf sputtered on sapphire with Pt and Zr as contact materials to form oxygen sensors. Ohmic contacts (Zr at low temperature, Pt at high temperatures) and Schottky barriers (Pt at low temperatures) were formed. The I–V curves showed large shifts of the forward-bias voltage upon exposure to oxygen (Fig. 3.73). The observed effect is analogous to the responses of the Pd–TiO_2 hydrogen sensor reported by Yamamoto et al. (1980). Because of the possible chemisorption of other gases, such as CO_x and NO_x, at the Pt–TiO_2 interface, this particular sensor seems to have a strong potential for developing several other sensors based on the detection principle of the variation of the Pt work function upon gas chemisorption at the metal–semiconductor interface. Petty (1986) has given a detailed discussion on the conduction mechanisms of Schottky diodes involving Pd–SiO_2–n-Si interfaces, in terms of interface states. This discussion appears to be relevant to work-function-shift-based sensors due to chemisorption, with some modification in the limit of no interfacial insulator layer, and may serve as a starting point toward a unified theory of Schottky-barrier-based gas sensor device physics.

REFERENCES

Abdullayev, A.G., Evdokimov, A.V., Murshudly, M.N., and Scheglov, M.I. (1987). *Sens. Actuators* **11**, 339.

Ackelid, U., Armgarth, M., Spetz, A., and Lundström, I. (1985). *IEEE Electron Device Lett.* **EDL-7**, 353.

Ackelid, U., Winquist, F., and Lundström, I. (1986). *Proc. Int. Meet. Chem. Sens., 2nd*, Bordeaux, Fr., p. 395.

Ackelid, U., Petersson, L.-G., and Lundström, I. (1987). *Europhys. Conf. Abstr., Eur. Phys. Soc.*, Luzern, Abstr. No. Pa 060.

Armgarth, M., and Nylander, C. (1981). *Appl. Phys. Lett.* **32**, 91.

Armgarth, M., and Nylander, C. (1982). *IEEE Electron Device Lett.* **EDL-3**, 384.

Armgarth, M., Nylander, C., Svensson, C., and Lundström, I. (1984). *J. Appl. Phys.* **56**, 2956.

Armgarth, M., Hua, T.H., and Lundström, I. (1985). *Dig. Tech. Pap. Transducers '85*, IEEE Cat. No. 8SCH2127-9. Philadelphia, p. 235.

Aspnes, D.E., and Heller, A. (1983). *J. Vac. Sci. Technol.* **B1**, 602.

Baidyaroy, S., and Mark, P. (1972). *Surf. Sci.* **30**, 53.

Behm, R.J., Penka, V., Cattania, M.-G., Christmann, K., and Ertl, G. (1983). *J. Chem. Phys.* **78**, 7486.

Bergveld, P. (1985). *Sens. Actuators* **8**, 109.

Bergveld, P., and van der Schoot, B.H. (1988). *Sel. Electrode Rev.* **10**, 5.

Bethe, H.A. (1942). *MIT Radiat. Lab. Rep.* **43–12**.

Bianchi, D., Garder, G.E.E., Pajonk, G.M., and Teichner, S.J. (1975). *J. Catal.* **38**, 135.

Blackburn, G.F., Levy, M., and Janata, J. (1983). *Appl. Phys. Lett.* **43**, 700.

Bower, R.W., and Mayer, J.W. (1972). *Appl. Phys. Lett.* **20**, 359.

Brews, J.R. (1977). *Solid-State Electron.* **20**, 607.

Brews, J.R. (1981). In *Applied Solid State Science* (D. Kahng, Ed.), Suppl. 2A. Academic Press, New York.

Brillson, L.J. (1977). *Surf. Sci.* **69**, 62.

Brillson, L.J. (1978). *Phys. Rev. Lett.* **40**, 260.

Burshtein, Z., and Levinson, J. (1975). *Phys. Rev.* **B12**, 3453.

Butler, M.A. (1985). *J. Appl. Phys.* **58**, 2044.

Card, H.C., and Rhoderick, E.G. (1971). *J. Phys. D.* **4**, 1589.

Choi, S.-Y., Takahashi, K., and Matsuo, T. (1984). *IEEE Electron Device Lett.* **EDL-5**, 14.

Choi, S.-Y., Takahashi, K., Esashi, M., and Matsuo, T. (1986). *Sens. Actuators* **9**, 353.

Corker, G.A., and Svensson, C.M. (1978). *J. Electrochem. Soc.* **125**, 1881.

Cowley, A.M., and Sze, S.M. (1965). *J. Appl. Phys.* **36**, 3212.

Cox, P.F. (1974). U.S. Patent 3,831,432.

Crowell, C.R., and Sze, S.M. (1966). *Solid-State Electron.* **9**, 1035.

Crowell, C.R., Sarace, J.C., and Sze, S.M. (1965). *Trans. Metall. Soc. AIME* **233**, 478.

Dalby, D., and Shurmer, H.V. (1988). In *Final Programme, Eurosensors II*, p. 147. Enschede, The Netherlands.

D'Amico, A., and Verona, E. (1988). *Prog. Solid State Chem.* **18**, 173.

D'Amico, A., Fortunato, G., Petrocco, G., and Coluzza, G. (1983a). *Appl. Phys. Lett.* **42**, 964.

D'Amico, A., Fortunato, G., Petrocco, G., and Coluzza, C. (1983b). *Sens. Actuators* **4**, 349.

Dannetun, H.M. (1967). Ph.D. Dissertation No. 160. University of Linköping, Sweden.

Dannetun, H.M., Lundström, I., and Petersson, L.-G. (1988a). *J. Appl. Phys.* **63**, 207.

Dannetun, H.M., Lundström, I., and Petersson, L.-G. (1988b). *Surf. Sci.* **193**, 109.

Deal, B.E. (1974). *J. Electrochem. Soc.* **121**, 198C.

Deal, B.E. (1980). *IEEE Trans. Electron Devices* **ED-27**, 606.

De Rooij, N.F., and Bergveld, P. (1978). *Phys. SiO_2 and Its Interfaces, Proc. Int. Top. Conf.*, Yorktown Heights, NY, *1978*, p. 433.

Dobos, K., Strotman, R., and Zimmer, G. (1983). *Sens. Actuators* **4**, 593.

Dobos, K., Armgarth, M., Zimmer, G., and Lundström, I. (1984). *IEEE Trans. Electron Devices* **ED-31**, 508.

Dobos, K., Mokwa, W., Zhang, Y., Dura, H.-G., and Zimmer, G. (1990). *Sens. Actuators* **B1**, 25.

Duszak, M., Jakubowski, A., and Sekulski, W. (1981). *Thin Solid Films* **75**, 379.

Engelhardt, H.A., Feulner, P., Pfnür, H., and Menzel, D. (1977). *J. Phys. E* **10**, 1133.

Evans, N.J., Petty, M.C., and Roberts, G.G. (1986). *Sens. Actuators* **9**, 165.

Fare, T.J., and Zemel, J.N. (1987). *Sens. Actuators* **11**, 101.

Fare, T.L., Lundström, I., Zemel, J.N., and Feygenson, A. (1986). *Appl. Phys. Lett.* **48**, 632.

Fare, T., Spetz, A., Armgarth, M., and Lundström, I. (1988). *Sens. Actuators* **14**, 369.

Fogelberg, J., Lundström, I., and Petersson, L.-G. (1987). *Phys. Scr.* **35**, 702.

Fonash, S.J., Huston, H., and Ashok, S. (1982). *Sens. Actuators* **2**, 363.

Ford Motor Co. (1982). European Patent Specification G001-510.

Fortunato, G., D'Amico, A., Coluzza, C., Sette, F., Capasso, C., Patella, F., Quaresima, C., and Perfetti, P. (1984). *Appl. Phys. Lett.* **44**, 887.

Fortunato, G., Bearzotti, A., Caliendo, C., and D'Amico, A. (1989). *Sens. Actuators* **16**, 43.

Fukui, K., and Komatsu, K. (1983). *Anal. Chem. Symp. Ser.* **17**, 52.

Gall, M., and Müller, R. (1988). In *Final Programme, Eurosensors II*, p. 216. Enschede, The Netherlands.

Goodman, A.M. (1963). *J. App. Phys.* **34**, 329.

Grove, A.S., Deal, B.E., Snow, E.H., and Sah, C.T. (1965). *Solid-State Electron.* **8**, 145.

Hagen, W., Lambrich, R.E., and Lagois, J. (1983). *Adv. Solid State Phys.* **23**, 259.

Heiland, G. (1954). *Z. Phys.* **138**, 459.

Heiland, G. (1982). *Sens. Actuators* **2**, 343.

Henisch, H.K. (1957). *Rectifying Semiconductor Contacts*. Oxford University Press (Clarendon), Oxford.

Hierold, C., and Müller, R. (1988). In *Final Programme, Eurosensors II*, p. 217. Enschede, The Netherlands.

Hornik, W. (1990). *Sens. Actuators* **B1**, 35.

Hua, T.H., Armgarth, M., and Lundström, I. (1984). *Proc. Nord. Semicond. Meet., 11th, 1984*, p. 229.

Hughes, R.C., Schubert, W.K., Zipperian, T.E., Rodriguez, J.L., and Plut, T.A. (1987). *J. Appl. Phys.* **62**, 1074.

Hunter, D.N., and Brooke, R.J. (1981). U.K. Patent Application 2,071,069A.

Ito, K. (1979). *Surf. Sci.* **86**, 345.

Ito, K., and Kojima, K. (1982). *Int. J. Hydrogen Energy* **7**, 495.

Jelly, K.W., and Maclay, G.J. (1987). *IEEE Trans. Electron Devices* **ED-34**, 2086.

Jeppson, K.O., and Svensson, C.M. (1977). *J. Appl. Phys.* **48**, 2004.

Josowicz, M., and Janata, J. (1986). *Anal. Chem.*, **58**, 514.

Kahng, D. (1976). *IEEE Trans. Electron Devices* **ED-23**, 655.

Keramati, B. (1980). Ph.D. Dissertation, University of Pennsylvania, Philadelphia.

Keramati, B., and Zemel, J.N. (1982a). *J. Appl. Phys.* **53**, 1091.

Keramati, B., and Zemel, J.N. (1982b). *J. Appl. Phys.* **53**, 1100.

Kirner, U., Schierbaum, K.D., Göpel, W., Leibold, B., Nicoloso,. N., Weppner, W., Fischer, D., and Chu, W.F. (1990). *Sens. Actuators* **B1**, 103.

Kittel, C. (1976). *Introduction to Solid State Physics*, 5th ed. Wiley, New York.

Kofstad, P. (1972). *Non-stoichiometry, Diffusion and Electrical Conductivity in Primary Metal Oxides.* Wiley (Interscience), New York.

Komori, N., Sakai, S., and Komatsu, K. (1987). *Proc. Int. Conf. Solid-State Sens. Actuators, Transducers '87, 4th*, Tokyo, p. 591.

Krey, D., Dobos, K., and Zimmer, G. (1983). *Sens. Actuators* **1**, 169.

Kricher, C.J. (1971). *Solid-State Electron.* **14**, 507.

Kroger, H., and Wegener, H.A.R. (1973). *Appl. Phys. Lett.* **23**, 397.

Kurtin, S., McGill, T.C., and Mead, C.A. (1969). *Phys. Rev. Lett.* **22**, 1433.

Lalauze, R., Gillard, P., and Pijolat, C. (1988). *Sens. Actuators* **14**, 243.

Lang, D.V. (1974). *J. Appl. Phys.* **45**, 3023.

Lenac, Z., Sunjič, M, Conrad, H., and Kordesch, M.E. (1987). *Phys. Rev. BII* **36**, 9500.

Lepselter, M.P., and Sze, S.M. (1968). *Bell Syst. Tech. J.* **47**, 195.

Lewis, F.A. (1967). *The palladium-Hydrogen System.* Academic Press, London.

Lindmeyer, J. (1965). *Solid-State Electron.* **8**, 523.

Logothetis, E.M., and Hetrick, R.E. (1979). *Solid State Commun.* **31**, 167.

Logothetis, E.M., Park, K., Meitzler, A.H., and Land, K.R. (1975). *Appl. Phys. Lett.* **28**, 209.

Lundström, I. (1981). *Sens. Actuators* **1**, 403.

Lundström, I., and Söderberg, D. (1981/82). *Sens. Actuators* **2**, 105.

Lundström, I., and Svensson, C. (1985). In *Solid State Chemical Sensors* (J. Janata and R.J. Huber, Eds.), Chapter 1. Academic Press, Orlando, FL.

Lundström, I., Shivaraman, M.S., and Svensson, C. (1975a). *J. Appl. Phys.* **46**, 3876.

Lundström, I., Shivaraman, M.S., Svensson, C.S., and Lundkvist, L. (1975b). *Appl. Phys. Lett.* **26**, 55.

Lundström, I., Shivaraman, M.S., and Svensson, C. (1977). *Surf. Sci.* **64**, 497.

Lundström, I., Armgarth, M., Spetz, A., and Winquist, F. (1986a). *Sens. Actuators* **10**, 399.

Lundström, I., Armgarth, M., Spetz, A., and Winquist, F. (1986b). *Proc. Int. Meet. Chem. Sens. 2nd*, Bordeaux, Fr., p. 387.

Lundström, I., Armgarth, M., and Petersson, L.-G. (1989). *CRC Crit. Rev. Solid State Mater. Sci.* **15**, 201.

Lundström, I., Spetz, A., Winquist, F., Ackelid, U., and Sundgren, H. (1990). *Sens. Actuators* **B1**, 15.

Maclay, G.J., Jelley, K.W., Nowroozi-Esfahani, S., and Formosa, M. (1988). *Sens. Actuators* **14**, 331.

Mandelung, O. (1978). *Introduction to Solid State Theory.* Springer-Verlag, Berlin.

Mariucci, L., Pecora, A., Puglia, C., Reita, C., Petrocco, G., and Fortunato, G. (1990). *Jpn. J. Appl. Phys.* **29**, L2357.

Marucco, J.F., Gautron, J., and Lemasson, P. (1981). *J. Phys. Chem. Solids* **42**, 363.

McAleer, J.F., Moseley, P.T., Norris, J.O.W., Williams, D.E., and Tofield, B.C. (1986). *Proc. Int. Meet. Chem. Sens.*, 2nd, Bordeaux, Fr., p. 264.

Miyahara, Y., Tsukada, K., and Miyagi, H. (1988). *J. Appl. Phys.* **63**, 2431.

Mokwa, W., Dobos, K., and Zimmer, G. (1987). *Sens. Actuators* **12**, 333.

Morrison, S.R. (1987). *Sens. Actuators* **12**, 425.

Müller, R., and Lange, E. (1986). *Sens. Actuators* **9**, 39.

Murakami, K., Ye, D.-B., and Yamamoto, T. (1988). *Sens. Actuators* **13**, 315.

Nicollian, E.H., and Brews, J.R. (1982). *MOS Physics and Technology.* Wiley, New York.

Nie, H.-Y., and Nannichi, Y. (1991). *Jpn. J. Appl. Phys.* **30**, 906.

Nylander, C., Armgarth, M., and Svensson, C. (1984). *J. Appl. Phys.* **56**, 1177.

Ogita, M., Ye, D.-B., Kawamura, K., and Yamamoto, T. (1986). *Sens. Actuators* **9**, 157.

Okada, D.N., Chun, C.K.Y., Yoshimi, L.K., and Holm-Kennedy, J.W. (1987). *Proc. Int. Conf. Solid-State Sens. Actuators, Transducers '87, 4th,* Tokyo, p. 561.

Park, K., and Logothetis, E.M. (1977). *J. Electrochem. Soc.* **124**, 1443.

Peschke, M., Lorenz, H., Riess, H., and Eisele, I. (1990). *Sens. Actuators* **B1**, 21.

Petersson, L.-G., Dannetun, H.M., and Lundström, I. (1984). *Phys. Rev. Lett.* **52**, 1806.

Petersson, L.-G., Dannetun, H.M., Fogelberg, J., and Lundström, I. (1985). *J. Appl. Phys.* **58**, 404.

Petersson, L.-G., Fogelberg, J., Dannetun, H.M., and Lundström, I. (1987a). *J. Vac. Sci. Technol.* **A5**, 819.

Petersson, L.-G., Dannetun, H.M. Fogelberg, J., and Lundström, I. (1987b). *Appl. Surf. Sci.* **27**, 275.

Petty, M.C. (1982). *Electron. Lett.* **18**, 314.

Petty, M.C. (1986). *Solid-State Electron.* **29**, 89.

Pietrucha, B., and Lalević, B. (1988). *Sens. Actuators* **13**, 275.

Plihal, M. (1977). *Siemens Forsch.- Entwicklungsber.* **6**, 53.

Poteat, T.L., and Lalević, B. (1981). *IEEE Electron Device Lett.* **EDL-2**, 82.

Poteat, T.L., and Lalević, B. (1982). *IEEE Trans. Electron Devices* **ED-29**, 123.

Poteat, T.L., Lalević, B., Kuliyer, B., Yousuf, M., and Chen, M. (1983). *J. Electron. Mater.* **12**, 181.

Rhoderick, E.H. (1978). *Metal-Semiconductor Contacts*. Oxford Univ. Press. (Clarendon), Oxford.

Rideout, V.L. (1978). *Thin Solid Films* **48**, 261.

Ross, J.F., Robins, I., and Webb, B.C. (1987). *Sens. Actuators* **11**, 73.

Ruths, P.F., Ashok, S., Fonash, S.J., and Ruths, J.M. (1981). *IEEE Trans. Electron Devices* **ED-28**, 1003.

Ryder, R.M. (1968). U.S. Patent 3, 360, 851.

Rye, R.R., and Ricco, A.J. (1987). *J. Appl. Phys.* **62**, 1084.

Saaman, A.A., and Bergveld, P. (1985). *Sens. Actuators* **7**, 75.

Sancier, K.M. (1971). *J. Catal.* **20**, 106.

Sekido, S., and Ariga, K. (1982). U.S. Patent Specification 4,314,996.

Senturia, S.D., Secken, C.M., and Wishnenski, J.A. (1977). *Appl. Phys. Lett.* **30**, 106.

Sermon, P.A., and Bond, G.C. (1973). *Catal. Rev.* **8**, 211.

Shivaraman, M.S. (1976). *J. Appl. Phys.* **47**, 3592.

Shivaraman, M.S., and Svensson, C. (1976). *J. Electrochem. Soc.* **123**, 1258.

Shivaraman, M.S., Lundström, L., Svensson, C., and Hammarsten, H. (1976). *Electron. Lett.* **12**, 483.

Sigrist, M. (1992). In *Progress in Photothermal and Photoacoustic Science and Technology* (A. Mandelis, Ed.), Chapter 7. Elsevier, New York.

Singh, J., and Madhukar, A. (1981). *J. Vac. Sci. Technol.* **19**, 437.

Spetz, A., Armgarth, M., and Lundström, I. (1987). *Sens. Actuators* **11**, 349.

Spetz, A., Armgarth, M., and Lundström, I. (1988). *J. Appl. Phys.* **64**, 1274.

Steele, M.C., and MacIver, B.A. (1976). *Appl. Phys. Lett.* **28**, 687.

Steele, M.C., Hile, J.W., and MacIver, B.A. (1976). *J. Appl. Phys.* **47**, 2537.

Stenberg, M., and Dahlenbäck, B.I. (1983). *Sens. Actuators* **4**, 273.

Stiblert, L., and Svensson, C. (1975). *Rev. Sci. Instrum.* **46**, 1206.

Stoneham, A.M. (1987). In *Solid State Gas Sensors* (P.T. Moseley and B.C. Tofield, Eds.), Chapter 8. Adam Hilger, Bristol.

Strassler, S., and Reis, A. (1983). *Sens. Actuators* **4**, 465.

Sullivan, T.E., Childs, R.B., Ruths, J.M., and Fonash, S.J. (1978). *Phys. SiO_2 and Its Interfaces, Proc. Int. Top. Conf.*, Yorktown Heights, NY, *1978*, p. 454.

Sundgren, H., Lundström, I., Winquist, F., Lukkari, I., Carlsson, R., and World, S. (1990). *Sens. Actuators* **B2**, 115.

Sze, S.M. (1981). *Physics of Semiconductor Devices*, 2nd ed. Wiley, New York.

Sze, S.M. (1983). *VLSI Technology*, McGraw-Hill Series in Electrical Engineering, Electronics and Electronic Circuits. McGraw-Hill, New York.

Taguchi, N. (1970). U.K. Patent Specification 1,280,809.

Thomas, J.M., and Thomas, W.J. (1967). *Introduction to the Principles of Heterogeneous Catalysis*, Chapter 2. Academic Press, New York.

Thorstensen, B. (1981). Thesis, Rep. STF 44A 81118. University of Trondheim, Norway.

Tien, T.Y., Stadler, N.L., Gibbons, E.F., and Zacmanidis, P. (1975). *Am. Ceram. Soc. Bull.* **54**, 280.

Wangsness, R.K. (1986). *Electromagnetic Fields.* Wiley, New York.

Watson, J. (1984). *Sens. Actuators* **5**, 29.

Weimer, P.K. (1962). *Proc. IRE* **50**, 1462.

Wen, C.C., and Zemel, J.N. (1979). *IEEE Trans. Electron. Devices* **ED-26**, 1945.

Williams, D.E. (1987). In *Solid State Gas Sensors* (P.T. Moseley and B.C. Tofield, Eds.), Chapter 5. Adam Hilger, Bristol.

Williams, D.E., Tofield, B.C., and McGeehin, P. (1984). U.S. Patent Specification 4,454,494.

Windischmann, H., and Mark, P. (1979). *J. Electrochem. Soc.* **126**, 627.

Winquist, F., Spetz, A., Armgarth, M., Nylander, C., and Lundström, I. (1983). *Appl. Phys. Lett.* **43**, 839.

Winquist, F., Spetz, A., Armgarth, M., Lundstöm, I., and Danielsson, B. (1985a). *Sens. Actuators* **8**, 91.

Winquist, F., Lundström, I., and Danielsson, B. (1985b). *Dig. Tech. Pap., Transducers '85*, IEEE Cat. No. 8SCH2127-9, Philadelphia, p. 162.

Wohltjen, H., Barger, W.R., Snow, A.W., and Jarvis, N.L. (1985). *IEEE Trans. Electron Devices* **ED-32**, 1170.

Wronsky, C.R., Carlson, D.E., and Daniel, R.E. (1976). *Appl. Phys. Lett.* **29**, 602.

Yamamoto, T., and Morimoto, M. (1972). *Appl. Phys. Lett.* **20**, 269.

Yamamoto, N., Tonomura, S., Matsuoka, T., and Tsubomara, H. (1980). *Surf. Sci.* **92**, 400.

Zaromb, S., and Stetter, J.R. (1984). *Sens. Actuators* **6**, 225.

Zemel, J.N. (1975). *Thin Solid Films* **28**, L5.

Zemel, J.N. (1981). *Sens. Actuators* **1**, 31.

Zemel, J.N., Young J.J., and Rahnamai, H. (1975). *CRC Crit. Rev. Solid State Science* **1**, 1.

Zemel, J.N., Keramati, B., Spivak, C.W., and D'Amino, A. (1981). *Sens. Actuators* **1**, 427.

Zhang, W., and Zhao, Y. (1988). *Sens. Actuators* **15**, 85.

Zimmer G., Höflinger, G., and Schneider, J. (1979). *IEEE J. Solid-State Circuits* **SC-14**, 312.

CHAPTER

4

PHOTONIC AND PHOTOACOUSTIC GAS SENSORS

4.1. INTRODUCTION

In recent years novel spectroscopic instrumentation and analytical techniques for gas sensing have appeared. The sensors fabricated thus far, based on spectroscopic principles, are mainly of two broad categories: (a) *photonic* (mainly optical, including reflectance, transmittance, absorptance, ellipsometric, fluorescence, plasmon resonance, and photoelectron spectroscopies); and (b) *photoacoustic* (mainly microphone–gas cell detection spectroscopies). A third and equally important category of photonic gas sensors includes fiber- and integrated-optic sensors and is discussed in Chapter 5.

4.2. PHOTONIC GAS SENSORS

In this family of sensors two signal generation schemes are possible: (a) *broadband excitation sources* (e.g., dispersive spectral lamps, blackbody radiators, Fourier transform infrared spectrometers); and (b) *narrowband coherent sources* (e.g., lasers). Owing to widely varying instrumentation techniques used in this rather young and still emerging subfield of gas sensor technology, no coherent literature exists and the field consists mainly of isolated attempts by a few investigating groups. A major advantage of the spectroscopic techniques, in comparison with nonspectroscopic detection, is in the ability of the former to couple intimately to the optical constants of the irradiated solid interface. Thus unique, unparalleled, and detailed information about the energetic structure of the chemical catalyst with or without the absorbed gaseous impurities can be obtained in principle at the electronic or molecular level. An additional advantage stems from the largely noncontact, remote-sensing character of spectroscopic sensors.

This feature tends to be partly offset by the usually complex and relatively sophisticated instrumentation and measurement systems required. In this section we focus on a few selected photonic gas sensors that are considered representative of broad spectroscopic detection principles and categories of experimental setups.

133

4.2.1.   Reflection Spectroscopic Sensors

One of the simplest optical principles is related to reflectance spectroscopy. Fortunato et al. (1989) applied this principle for the first time to monitor the modifications of the electronic structure of Pd film gates of MOS capacitors during hydrogen gas absorption and desorption. As a complementary measurement channel they monitored the electrical conductance of the Pd films, a conventional technique with well-understood signal interpretation (Lundström, 1981). The reflectance measurements were performed at an angle of incidence of 45°. The light from the tungsten filament spectral lamp was filtered by a Monospec 600 monochromator, and the relative reflectance from the Pd–MOS capacitors was detected by using a silicon photodiode in the wavelength range 0.4–1.1 μm and by a Golay cell in the infrared range up to 2.07 μm (0.6 eV). Optical intensity modulation and lock-in detection of the signal was used, resulting in an instrumental sensitivity to variations on the order of 0.1% in reflectance and 0.01% in conductance signals. Reflectance variations were simultaneously monitored at Pd-coated quartz substrate interfaces, and the observed changes were consistent with rapid diffusion of hydrogen gas from the exposed external surface to the Pd–quartz

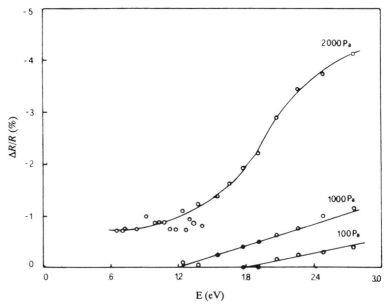

Figure 4.1. The relative reflectance, $\Delta R/R$, vs. the photon energy, E, for three different hydrogen partial pressures: 2000, 1000, and 100 Pa (Fortunato et al., 1989).

interface. Based on this observation Fortunato et al. (1989) assumed that external reflectance changes were representative to bulk Pd-film modifications in the presence of ambient H_2 gas. Figure 4.1 shows relative reflectance dependences of the sensor on excitation optical beam photon energy, under three different hydrogen partial pressure, $p(H_2)$, values. The spectrum of $\Delta R/R$ decreases with decreasing photon energy. This downward trend was interpreted by the dependence of the reflectance on the position of the Fermi level on the α-phase Pd metal. The Fermi level is known to rise with hydrogen concentration (Wicke and Brodowsky, 1978), moving on the tail of the $4d$ electrons. As the Fermi level rises it causes a reduction in the number density of free states just above the Fermi level, thus reducing the number of optical transitions. At higher $p(H_2)$ (> 1000 Pa), the influence of the α–β transition on the band structure of the Pd film may be responsible for the nonlinear features observed in the $p(H_2) = 2000$ Pa curve in Fig. 4.1.

Time responses of the reflectance, the conductance, and the flat-band voltage shift of the sensor enabled investigators to gain further understanding of the Fermi level shift with hydrogen concentration: two contributions were identified consistent with two distinctly different adsorption sites in Pd, with relative coverages exhibiting Langmuirian behavior. Figure 4.2 shows the energy level scheme proposed by Fortunato et al. (1989). An electrical dipole effect can be associated with the hydrogen adsorbed on sites of the first type, whereas the hydrogen adsorbed on sites of the second type produces a shift of the Fermi level, altering the flat-band voltage through the variation of the metal–semiconductor work function difference. Further support for the

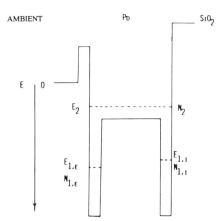

Figure 4.2. Energy level scheme for hydrogen atoms in the Pd–SiO$_2$ system: external surface sites $N_{1,E}$ with relative energy $E_{1,E}$; interfacial sites $N_{1,I}$ with relative energy $E_{1,I}$; and bulk sites N_2 with relative energy E_2 (Fortunato et al., 1989).

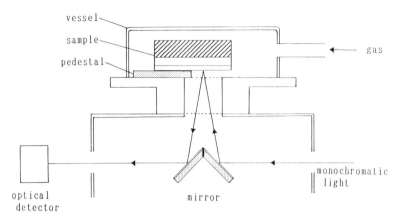

Figure 4.3. Schematic diagram of the hydrogen sensor assembly developed by Ito and Oghami (1992). The near-normal reflectance of a sample is measured by a photomultiplier or PdS cell as a function of hydrogen concentration in the vessel.

proposed model of the flat-band voltage shift was given by the two distinct features observed in transient current experiments carried out at room temperature (Reihua et al., 1985). At this time it appears that reflectance-based hydrogen sensors can be powerful probes of the electronic structure of palladium (Frazier and Glosser, 1982).

A recent variant of this technique is the hydrogen sensor developed by Ito and Ohgami (1992). These workers used a very thin Pd overlayer deposited on an anodically oxidized tungsten sheet. The near-normal reflectance of a sample was measured by a spectrophometer (Shimazu model UV-3100). Figure 4.3 shows the instrumental outline. Tungsten oxide sheet reduction by hydrogen molecules in the presence of a very thin catalytic Pd overlayer was found to occur reversibly at room temperature. The H_2 molecules are dissociated into hydrogen atoms on the catalyst surface, followed by diffusion into the interior of the oxide to form the tungsten blue phase. This is analogous to the H_2 effect on tungsten oxide powder particles, where the diffusion of the atomic (reducing) species forms hydrogen tungsten bronzes (blue coloration) (Kohn and Boudart, 1964). When exposed to hydrogen gas, the interaction induces a modulation in the film optical reflectance as shown in Fig. 4.4. This effect is similar to that observed by Fortunato et al. (1989) with their Pd-gate MOS capacitor. The intensity of the reflected light could be enhanced by tuning to wavelengths corresponding to a constructive interference fringe within the anodic oxide film. Upon exposure to H_2 gas the interference color of the film was changed, while the infrared reflectance decreased owing to an increase in the extinction coefficient of the film surface. The authors

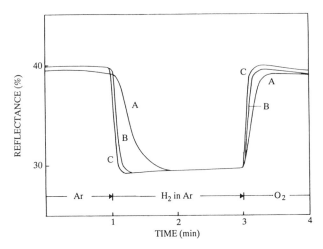

Figure 4.4. Transient of the reflectance at $\lambda = 1.4\,\mu$m. Curves A, B, and C correspond to transients when the hydrogen sensor was exposed to argon containing 0.1, 0.5, and 1% H_2, respectively. Flow rate: 300 cm³/min (Ito and Ohgami, 1992).

suggested on the basis of the simple Drude model (Goldner et al., 1983) that the hydrogen atoms, upon diffusion into the bulk of the Pd film, give rise to an increase of the free electron concentration in the film. This increase, in turn, enhances the optical absorption coefficient of the film, thus depressing the optical reflectance. A concomitant decrease in the refractive index of the anodic oxide was further observed, resulting in shifting the reflectance extrema in the visible region to shorter wavelengths.

Ito and Ohgami (1992) also studied hydrogen isotope effects with their sensor and found that its response to deuterium was 66% slower.

4.2.2. Spectroscopic Ellipsometry Sensor

Lundström's group has recently developed a gas sensor based on tetra-sulfonated copper phthalocyanine (CuTSPc) thin films ($\simeq 90$ Å) spin cast on gold (Martensson et al., 1990). It is found that the dielectric function of the films, as measured with spectroscopic ellipsometry, was sensitive to the presence of NO_2/N_2O_4 (NO_x) gases. Films of CuPc are stable, and the response time to NO_x is short compared to many other similar compounds. Thin films of CuTSPc were synthesized (Weber and Busch, 1965). The gold substrates on which these films were spun were made by sequential vacuum deposition of chromium (30 Å) and gold (1500 Å) onto spontaneously oxidized silicon. The chromium layer was inserted to promote adhesion between the gold and the oxidized silicon surface. A rotating spectroscopic ellipsometric analyzer operating in the photon energy range 1.5–4.5 eV (825–275 nm) was used to obtain spectra in air at room temperature (Aspnes and Studna, 1975). The ellipsometer measured the change in the state of polarization for the reflection of polarized light that was incident on the sample at an oblique angle. The measured quantity by such an instrument is the complex reflectance ratio

$$\rho = R_p/R_s = \tan(\psi)\exp(i\Delta) \tag{4.1}$$

where R_p and R_s are the complex reflectances of obliquely incident light polarized parallel and perpendicular to the plane of incidence, respectively; ψ and Δ are known as the "ellipsometric angles" (Assam and Bashara, 1977). The technique of spectroscopic ellipsometry allows the extraction of important information concerning the complex dielectric function of the film material

$$\epsilon = \epsilon_1 + i\epsilon_2 \tag{4.2}$$

Figure 4.5 shows the response ψ to pulses of NO_x gas, and Fig. 4.6 displays ϵ calculated by standard numerical inversion techniques using a three-layer model (ambient–film–substrate) and a 290 ppm NO_x in argon ambient. Both real and imaginary parts of ϵ are shown, calculated from the amplitude and phase data of the photomultiplier signal (Aspnes and Studna, 1975). The kinetic behavior shown in Fig. 4.6 (response to the first NO_x pulse) is similar to that obtained by Lloyd et al. (1988) using a surface plasmon resonance technique (see below). The slow part in the response to the third pulse (Fig. 4.6)

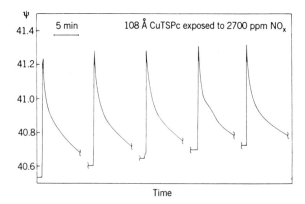

Figure 4.5. The ellipsometric response ψ of a 108 A CuTSPc film to 2700 ppm pulses of NO_x in argon with an interval of 30 min. The photon energy was 2.22 eV (Martensson et al., 1990).

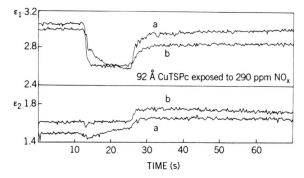

Figure 4.6. The ellipsometric response of the real (ϵ_1) and imaginary (ϵ_2) parts of the dielectric function of a 92 Å CuTSPc film to 290 ppm NO_x in argon. The photon energy was 2.0 eV. (a) Reponse to the first pulse. (b) Response to the third pulse (Martensson et al., 1990).

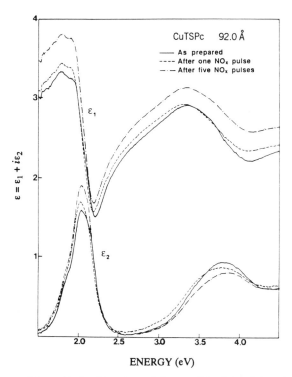

Figure 4.7. Spectra of the real (ϵ_1) and imaginary (ϵ_2) parts of the dielectric function of a CuTSPc film before and after exposure to pulses of 290 ppm of NO_x (Martensson et al., 1990).

was not observed in the experiments by Lloyd et al. (1988). Figure 4.7 shows ϵ_1 and ϵ_2 spectra before and after exposure to NO_x. The real part of ϵ is a measure of the density of dipoles, which increases when gas is absorbed. The imaginary part of ϵ is proportional to the optical absorption coefficient, α, of the CuTSPc. Figure 4.7 indicates that α increases after exposure to NO_x gas, in contradiction with results by Lloyd et al. (1988). The 2.0-eV-centered band (Q-band) is also slightly shifted to lower energies (red-shifted) after exposure.

The approximate detection of the spectroscopic ellipsometric gas sensor was found to be 100 ppm. Schoch and Temofonte (1988) used infrared (IR) spectroscopic detection to test phthalocyanine thin-film sensors in nitrogen dioxide and estimated an optical detection limit of 100–200 ppm. Their electrical detection limit was 1–10 ppb. The ellipsometric sensor developed by Martensson et al. (1990) is a potentially valuable new instrument yielding optical absorption as well as optical phase change information. Questions of gas selectivity and sensitivity still remain open, while further optimizations of the thickness of the CuTSPc and probe wavelength may render it more competitive with electronic (semiconductor-based) sensors (see Chapter 3).

4.2.3. Surface Plasmon Resonance Spectroscopic Sensors

The recent impetus on integration of organic layers, such as Langmuir–Blodgett films with microelectronic and optoelectronic devices, has led to the development of new optical spectroscopic gas sensors based on surface plasmon resonance (SPR) devices. In this type of device, changes in the optical properties of an active layer in response to external ambients are measured by monitoring the coupling of photons to surface plasmons at the interface between a metal and the active layer (Raether, 1977). It has been claimed that changes in the refractive index of the active layer on the order of 10^{-5} may be monitored by this method (Liedberg et al., 1983). Nylander et al. (1982) first described gas-sensing experiments using the Kretschmann configuration (see Kretschmann and Raether, 1968). A spin-coated active layer was used, consisting of a thin oil film of a silicon–glycol copolymer. This SPR instrument was able to detect anesthetic halothane gas in the 100 ppm to a few percent range. The estimated theoretical lower detection limit was 10 vpm (vapor parts per million). Lloyd et al. (1988) have reported the use of a silicon phthalocyanine Langmuir–Blodgett film as the basis for an SPR gas sensor and demonstrated its sensitivity to the detection of nitrogen dioxide. The equipment used has been described by Petty (1987): Microscope slides were rigorously cleaned by refluxing in isopropyl alcohol vapor and then coated with a silver layer ca. 50 nm thick deposited by vacuum evaporation. Subsequently, a thin layer ($\simeq 1$ nm) of nickel was evaporated onto

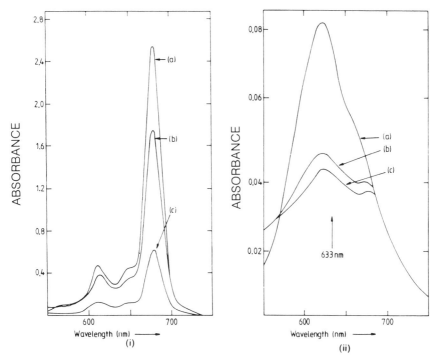

Figure 4.8. Optical absorption spectra showing effects of NO_x on CuPc: (i) in solution (chloroform concentration: 9.2×10^{-3} mg/mL); (ii) in Langmuir–Blodgett film form (five layers) on indium–tin oxide. See the text for an explanation of curves (a), (b), and (c) (Zhu et al., 1990).

the silver without breaking the vacuum. Layers of copper tetra-*tert*-butylphthalocyanine (CuPc) were deposited onto the metallized slides using the Langmuir–Blodgett technique. Optical absorption measurements were made with a Cary 2300 UV–VIS–IR spectrophotometer. Lloyd et al. (1988) reported significant and irreversible shifts of the SPR curve for the silver base layer on exposure to NO_x. This was attributed to the growth of a surface layer on the metal film. The same group (Zhu et al., 1990) reported a more effective protective coating by evaporating an ultrathin ($\simeq 1$ nm) layer of nickel onto the silver. The Ag/Ni combination produced no further change to the SPR reflectance vs. internal angle curve after a 25-min exposure to 1000 ppm NO_x and was therefore used as the base layer for experiments with the group's SPR apparatus. Figure 4.8 shows optical absorption spectra of CùPc as affected by exposure to NO_x, in both solution and Langmuir–Blodgett form. In the spectra of Fig. 4.8(i), curve (a) is for the unexposed solution; curve (b) was obtained immediately after the NO_x had passed through the solution

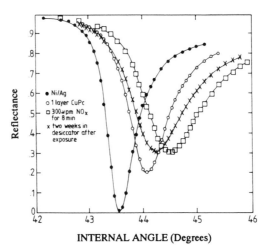

INTERNAL ANGLE (Degrees)

Figure 4.9. Effects on SPR curves at 633 nm of exposing one Langmuir–Blodgett layer of CuPc to 300 vpm (vapor parts per million) NO_x (Zhu et al., 1990).

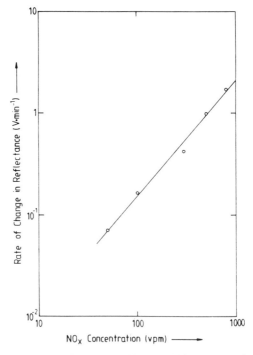

NO_x Concentration (vpm)

Figure 4.10. Rate of change of reflectivity at 633 nm vs. NO_x concentration (Zhu et al., 1990).

for 4 min; and curve (c) is for the same solution 12 h later, with no further exposure to nitrogen dioxide. The series of curves reveals that the dye solution is simply bleached by the oxidizing gas, and the process was found to be completely irreversible. Similar behavior was exhibited in the spectra of Fig. 4.8(ii), but the effect of the oxidizing gas on the optical absorption was found to be slowly reversible when five layers of CuPc were deposited onto conducting indium-tip oxide glass. Figure 4.9 shows the effect of NO_x exposure for 8 min on SPR curves for the CuPc Langmuir–Blodgett layers. The observed shift was partly and very slowly reversible. A more rapid albeit partial recovery could be obtained by heating the gas-exposed sample to 150 °C. Figure 4.10 shows that the rate of change of SPR reflectance is linear with NO_x concentration over the range studied. A lower detection limit for their SPR apparatus was estimated at a few vpm by Zhu et al. (1990).

SPR sensors appear to be promising in terms of sensitivity to some ambient gases, but a serious drawback remains as a result of their partial irreversibility. Furthermore, the selectivity has not yet been studied.

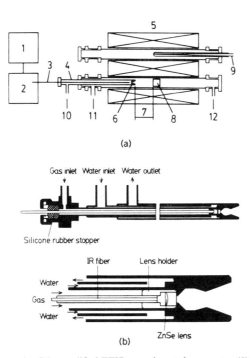

(a)

(b)

Figure 4.11. (a) Schematic of the modified FTIR experimental apparatus: (1) computer; (2) FTIR spectrometer; (3) IR fiber (KRS-5); (4) water-cooled probe; (5) resistance furnace; (6) ZnSe lens; (7) optical length; (8) alumina radiation plate; (9) thermocouple; (10) N_2 gas inlet; (11) gas outlet; (12) gas inlet. (b) Construction of the IR optical sensor (Maeda et al., 1990).

4.2.4. Infrared Spectroscopic Sensors

Very recently Maeda et al. (1990) adapted a Fourier transform infrared (FTIR) spectrometer to be used for in situ chemical gas sensing. Figure 4.11 shows a schematic of the apparatus. Their FTIR spectrometer, model JIR-100 (JEOL) uses a specially designed optical system for adapting an optical fiber as an external light source. The water-cooled probe, made of a triple copper shell, includes a zinc selenide lens and an infrared polycrystalline optical fiber of KRS-5 (TlI–TlBr) with transmission loss ca. 1 dB/m at 7 μm. The ZnSe lens focuses the IR radiation from the furnace onto the mirror polished edge of the fiber. The radiation is thus transferred into the spectrometer to produce an IR absorption spectrum of the gases present in the furnace.

The maximum sensitivity of the in situ determination apparatus was examined using CO gas. Figure 4.12(a) shows spectra obtained for partial pressure in the 2–100% range. Figure 4.12(b) demonstrates the ability of the instrument to detect 100 ppm ambient CO gas at a temperature of 1550 °C

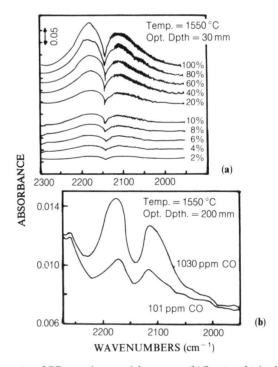

Figure 4.12. (a) IR spectra of CO at various partial pressures. (b) Spectra obtained at low p(CO) (Maeda et al., 1990).

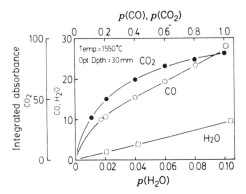

Figure 4.13. Calibration curves for CO, CO_2, and H_2O gases at 1550 °C (Maeda et al., 1990).

and an optical path length ("depth") of 200 mm. The same apparatus was used to detect CO, CO_2, and H_2O gases. The integrated absorptance was calculated from the area of the appropriate absorption band above the baseline for each component around $2140\,cm^{-1}$ (CO), $2300\,cm^{-1}$ (CO_2), and $1600\,cm^{-1}$ (H_2O). Calibration curves were thus obtained and are shown in Fig. 4.13. In addition, the change in partial pressure of CO gas was monitored in situ as it evolved following the reduction of SiO_2 by graphite according to the following equations:

$$SiO_2(s) + C(s) \rightarrow SiO(g) + CO(g) \tag{4.3a}$$

$$SiO_2(s) + 3C(s) \rightarrow SiC(s) + 2CO(g) \tag{4.3b}$$

A new and interesting sensor based on the spontaneous Raman scattering of laser light (Adler-Golden et al., 1990) was developed by Adler-Golden et al. (1992). In spontaneous Raman scattering, a small portion of the incident radiation from a laser or another optical source is scattered by a molecule at a characteristic red-shifted (Stokes) wavelength that corresponds to the loss of a vibrational quantum of energy. A spectrally selective device is used to discriminate the shifted Raman-scattered, light from the stray (or Rayleigh-scattered) laser light, the background fluorescence, and the Raman signals from other species. The intensity of the Raman scattering is proportional to the intensity of the illumination, the Raman cross section of the molecule, and its number density in the sample volume. The top and side view of the prototype instrument used to detect hydrogen gas is shown in Fig. 4.14. Several experimental runs of the hydrogen detection instrument were performed in the range 400 to 10^6 ppm to pure H_2 in background N_2 and He gases as well as in compressed and ambient air. The optical scattering signal

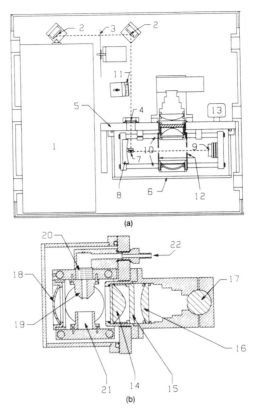

Figure 4.14. (a) Top cross-sectional view of the prototype hydrogen detector: (1) laser head; (2) steering mirrors; (3) chopper wheel; (4) lens; (5) bulkhead; (6) sample chamber cover; (7) reflective prism; (8) front cavity mirror; (9) rear cavity mirror; (10) support rails; (11) photodiode; (12) baffles; (13) pressure transducer. The laser-beam path is depicted by the dotted line. Only the first pass in the sample chamber is shown. (b) Side cross-sectional view of the sample chamber and the imaging optics: (14) collection lens; (15) interference filter; (16) focusing lens; (17) photomultiplier tube; (18) concave mirror; (19) flow injector; (20) flow straightener; (21) gas deflector; (22) gas inlet (Adler-Golden et al., 1992)

was found to be proportional to the H_2 concentration and independent of the background gas composition (Fig. 4.15). Room dust particles gave a flow-rate-dependent signal equivalent to a few hundred ppm of H_2. In addition, a small flow-rate-independent signal persisted at the level of ca. 2% $[H_2]$ at typical room humidities. This signal component was ascribed to Raman scattering from atmospheric water vapor. An increase in H_2 sensitivity with an increase in laser power was observed, as well as with an increase in photon energy, owing to the enhanced Raman cross section and the size and

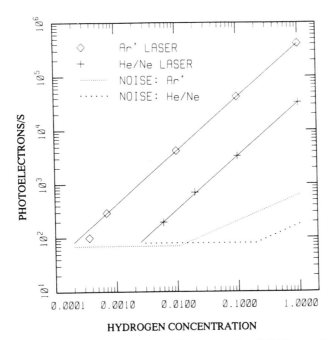

Figure 4.15. Dependence of signal and noise on H_2 concentration. Solid lines are based on the 100% H_2 values, assuming proportional instrument response. The 0.01 and 0.10 concentration points are from certified standard gas mixtures. Dotted curves are measured rms (root-mean-square) noise levels with a 2 s time constant (Adler-Golden et al., 1992).

quantum efficiency of the PMT (photomultiplier tube) photocathode. These instrumental optimizations increased the sensitivity with an Ar-ion laser to below 20 ppm H_2.

4.2.5. Photoemission Spectroscopic Sensor

Fortunato et al. (1984) used photoemission spectroscopy with synchrotron radiation to probe the mechanism of the Pd/SiO$_x$/a-Si:H hydrogen sensor. The conventional photoemission spectra approximately reproduce the features of the density of occupied states, shifted by the photon energy $h\nu$. The photoelectron distribution upper edge of the clean substrate corresponds to the top of the valence band. For metal–semiconductor and semiconductor–semiconductor interfaces with thin overlayers, the spectra reflect the electronic states of both substrate and overlayer. The spectra also reveal any shifts of the substrate valence-band edge due to changes in the band bending (Margaritondo, 1983). Upon exposure to hydrogen gas of a suitable interface solid-state structure a modification in the photoemission spectrum occurs (Fig. 4.16), thus qualifying the photoemission instrument as a photonic gas sensor. The hydrogen-induced shift $\Delta E = 0.3$ eV toward higher binding energies of the photoelectrons was explained by the reduction in the contact potential of the MIS structure. This interpretation was supported by the fact that alternate exposures to hydrogen and oxygen caused reversible shifts from curve (b) to curve (c) of Fig. 4.16 and vice versa. The a-Si hydrogenation could be related to the presence of atomic hydrogen produced by an ionization filament, to a small Si film thickness, and/or to a high defect density. The 0.3 eV shift of the SiO_x valence-band edge was due to hydrogen absorbed by the Pd film through its free surface (see inset of Fig. 4.16). This explanation was confirmed by curve (c), which was obtained by exposing the hydrogenated system to 5×10^{-5} torr O_2 at 110 °C for 1 min. After that treatment the

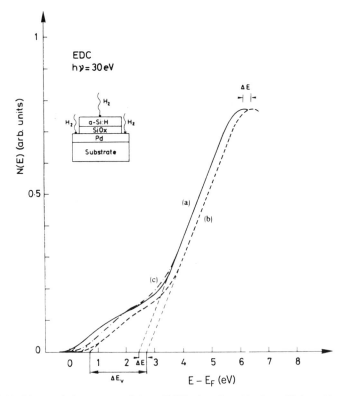

Figure 4.16. Photoemission spectra of the a-Si/SiO_x interface: (a) after a-Si deposition; (b) after hydrogen exposure; (c) after oxygen exposure. Here E_F is the Fermi level of the structure (Fortunato et al., 1984).

valence band shifted back rigidly and the SiO_x valence-band edge recovered the initial energy position. From the photoemission spectroscopic behavior shown in Fig. 4.16 it can be concluded that synchrotron sources can be used successfully with gas-sensing instrumentation to detect changes in the band bending of semiconductor-based structures due to the presence of hydrogen. Hydrogen sensitivity was found to be related to hydrogen-induced Schottky barrier modulation (see Fig. 3.63 in Chapter 3).

4.3. PHOTOACOUSTIC GAS SENSORS

The photoacoustic (PA) signal generation in gases has been described in detail by Kreuzer (1977) and Tam (1983). Most recently, Sigrist (1992) has reviewed applications of PA sensors to environmental and chemical trace gas analysis, mainly in multicomponent gases. Starting with a one-component gas buffered to atmospheric pressure by a nonabsorbing buffer gas, with a total molecular number density N and an absorption cross section σ, one can define the heat production rate H generated by absorption and optical-to-thermal energy conversion of modulated radiation of intensity I_0 thus (Sigrist, 1992):

$$H(\omega) = H_0 e^{i(\omega t - \phi)} \tag{4.4}$$

where

$$H_0 = \frac{N\sigma I_0}{[1 + (\omega\tau)^2]^{1/2}} \tag{4.5a}$$

and

$$\phi = \tan^{-1}(\omega\tau) \tag{4.5b}$$

Here $\omega = 2\pi f$, where f represents the intensity modulation frequency of the incident radiation; τ denotes the total lifetime of the optically excited molecular state (including radiative and nonradiative relaxation); and ϕ is the phase lag between heat production and incident radiation. In PA and PT (photothermal) detection schemes, optical saturation effects on these measurements have been studied by Bialkowski and Long (1987).

With respect to chemical trace gas analysis or air pollution studies, multi-component gas mixtures are usually encountered, which can be characterized by concentrations C_j and absorption cross sections σ_j of species j. The total heat production rate is then given by

$$H_0 = I_0 N_{tot} \sum_{j=1}^{n} C_j \sigma_j \tag{4.6}$$

where low modulation frequencies were assumed, such that $\omega\tau \ll 1$, as well as absence of optical saturation; N_{tot} denotes the total number density of molecules, and n represents the number of absorbing species. The relaxation lifetime τ_j is known to depend on gas concentration; therefore care must be taken in the application of Eq. (4.6) (Wolff and Peel, 1985).

The generation of acoustic waves can be calculated in relation to the heat production rate H. This problem has been studied by Kreuzer (1977) and Hess (1992) for nonresonant and resonant PA cells.

Since resonant spectrophones offer distinct advantages for trace gas detection compared to nonresonant cells, such as high sensitivity, range of modulation frequencies, and suppression of external noise, the present discussion will be restricted to the resonant type. For harmonic modulation of the incident light of intensity $I(r)$, the pressure variation amplitude $p(r, \omega)$ can be expressed in terms of a superposition of the normal modes $p_k(r)$ of the spectrophone (Kreuzer, 1977; Hess, 1983):

$$p(r, \omega) = \sum_k A_k(\omega)p_k(r) \qquad (4.7)$$

where

$$A_k(\omega) = a(\omega) \int_v p_k(r) H(r, \omega) \, dV \qquad (4.8)$$

Here V denotes the cell volume, and $H(r, \omega)$ is the heat production rate. The normal modes $p_k(r)$ depend strongly on the cell geometry. Moreover, the solutions $p(r, \omega)$ depend on the excitation geometry and the types of excited modes. For a cylindrical resonator operated at the first radial acoustic modes it can be shown (Kreuzer, 1977) that

$$p_1(r, t) = (\gamma - 1)(G/V)(Q_1/\omega_1)p_1(r)H_0 \exp[i(\omega_1 t - \phi)] \qquad (4.9)$$

where $\gamma = C_p/C_v$ is the ratio of the specific heats; G is a geometrical factor that depends on the transverse laser beam profile but not on the cell length; $p_1(r)$ denotes the normalized pressure amplitude of the first radial normal mode, related to the zeroth-order Bessel function J_0. The quality factor Q_1 is related to the damping of the first radial mode caused by heat conduction and viscosity losses of acoustic energy; Q_1 is also related to the full width $\Delta\omega_1$ of the resonant shape of $A_1(\omega)$ taken at $A_1(\omega)/\sqrt{2}$ by

$$Q_1 = \Delta\omega_1/\omega_1 \qquad (4.10)$$

The pressure amplitude is detected by one or more microphones placed at appropriate positions r_{mic}. Equation (4.6) can be modified to yield the microphone signal amplitude S for a mixture composed of n absorbing components:

$$S = CPN_{tot} \sum_{j=1}^{n} C_j \sigma_j \qquad (4.11)$$

where P is the average laser power and C is the cell constant. For cylindrical resonant cells, Eq. (4.9) gives:

$$C = R_{mic}(\gamma - 1)(G/V)(Q_1/\omega_1)p_1(r_{mic}) \qquad (4.12)$$

where R_{mic} is the microphone responsivity ($\sim 100\,mV/Pa$, condenser microphones; $\sim 10\,mV/Pa$, electret microphones). In some gas detection cases nonzero phase shifts have to be taken into account owing to relaxation mechanism effects, such as for air samples containing CO_2 that are investigated by CO_2 laser PA spectroscopy. This results in a corresponding phase shift ϕ for the microphone signal S. In these cases the look-in analyzer in-phase microphone signal is considered experimentally (Sigrist, 1992):

$$S \cos \phi = CPN_{tot} \left[\sum_{j=1}^{n-1} C_j \sigma_j - \left(\frac{\Delta E_{N_2'}(v_1)}{h v_{laser}} - 1 \right) C_{CO_2} \sigma_{CO_2} \right] \qquad (4.13)$$

where $\Delta E_{N_2'}(v_1)$ denotes the energy of the excited N_2 level in the vibrational mode v_1.

Equations (4.11) and (4.13) form the basis for the PA spectroscopic detection of gases. Sigrist (1992) presented a numerical technique in order to extract the n concentrations C_j from a set of m measured absorption coefficients α_i^ϵ for each of the m laser lines. More robust multivariate calibration approaches have also been suggested for use with PA spectroscopic gas sensors (Martens and Nase, 1989; Haaland, 1990). The choice and number of laser transitions considered for the multicomponent analysis strongly affect the accuracy of the concentrations determined (Frans and Harris, 1985). The detection *sensitivity* is defined as the minimum detectable concentration $C_{j,min}$ of gas j under interference-free conditions:

$$C_{j,min} = \frac{1}{N_{tot}} \min_i \left(\frac{\alpha_{min}}{\sigma_{ij}} \right) \qquad (4.14)$$

where α_{min} represents the minimum detectable absorption coefficient allowed by the instrument. Typically, $\alpha_{min} \sim 10^{-8}\,cm^{-1}$ in operating PA systems.

With $\sigma_{ij} \sim 10^{-19}\,cm^2$ and $N_{tot} \sim 10^{19}\,cm^{-3}$, one obtains $C_{j,min} \sim 10\,ppbv$. In monitoring ethylene exhalation of flowers, Bicanič et al. (1989) have achieved a detection threshold of 6 pptv. The detection *selectivity* of a specific gaseous species is determined by interference effects and can thus be influenced by a proper choice of the laser transitions used for the measurement (Meyer and Sigrist, 1990).

A typical setup for PA spectroscopic gas sensing studies is shown in Fig. 4.17. It consists of a tunable CW (continuous wave) laser source whose amplitude-modulated beam is directed through the PA cell containing the gas under investigation. The generated acoustic waves are detected by a microphone whose signal is fed to a lock-in amplifier locked to the modulation frequency. The average laser power P is measured simultaneously in order to normalize the microphone signal S, according to Eq. (4.11). Most photoacoustic studies have been performed with line-tunable CO and CO_2 lasers in the wavelength ranges 5–6 and 9–11 μm, respectively.

Since the pioneering works by Kerr and Atwood (1968) and by Kreuzer (1971), numerous laser PA studies of gaseous media have been reported. Table 4.1 gives a list of species detected and references. The first PA study on a multicomponent gas sample was devoted to the detection of NO in ambient air and in automobile exhausts with a spin-flip Raman (SRF) laser system (Kreuzer and Patel, 1971). The system allowed the detection of 10^8 NO molecules/cm³ corresponding to a concentration of < 1 ppbv. Recently, PA studies of air pollution monitoring have appeared, the first of this type being the CO-laser-induced analysis of motor vehicle exhausts, yielding the concentrations of 12 different compounds, mainly hydrocarbons (Bernegger and Sigrist, 1990; Bernegger et al., 1988; Sigrist et al., 1989b). The results of these studies are shown in Table 4.2. Further in situ investigations include the monitoring of ammonia (NH_3) in ambient air at ppbv concentrations with a Stark cell (Sauren et al., 1989), as well as with an automated mobile system using a conventional PA cell (Rooth et al., 1990). Olafsson et al. (1989), were able to detect ca. 1 ppmv NH_3 in the presence of high CO_2 concentrations by means of a waveguide CO_2 laser.

Figure 4.18 shows a simplified scheme of a dual-beam PA setup utilized by Sigrist (1992). The CO laser was tuned by a computer-controller grating with 150 lines/mm, blazed at 6.0 μm. The pyroelectric power detector was used as a fast-feedback generator to the computer. The chopper frequency could be adjusted to the resonance frequency of the PA cell (555 Hz) with an accuracy better than 0.05 Hz. The two beam paths through the air between the laser output and the entrance window of the PA cells were carefully matched to identical length in order to provide the same spectral composition of the laser beam at any line in both cells. Both sample and reference cells were of identical design (Bernegger and Sigrist, 1987, 1990) except that the

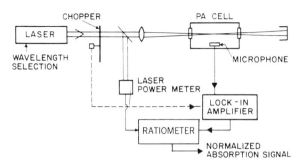

Figure 4.17. Typical experimental arrangement for laser PA spectroscopic detection of gases (Sigrist, 1992).

Table 4.1. Examples of PA Studies on Trace Gases in Samples of Ambient and Exhaust Air

Laser	Species	Source/Location	References
CO–SFR	NO	Ambient air, automobile exhaust	Kreuzer and Patel (1971)
CO–SFR	NO	Stratosphere, in situ	Patel et al. (1974); Burkhardt et al. (1975); Patel (1976)
CO_2	C_2H_4	Ambient air, fruit storage chamber	Perlmutter et al. (1979)
CO	Various hydrocarbons	Motor-vehicle exhausts	Bemegger et al. (1988); Sigrist et al. (1989a)
CO_2	C_2H_4	Ambient air, in situ	Meyer et al. (1988)
CO_2	Various hydrocarbons	Industrial exhausts	Sigrist et al. (1989b)
CO_2	NH_3	Inoculated beef	Bicanič et al. (1989)
CO_2	C_2H_4	Exhalation of flowers	Bicanič et al. (1989)
CO_2	NH_3	Ambient air	Sauren et al. (1989)
CO_2	NH_3	Ambient air	Rooth et al. (1990)
CO_2 waveguide	NH_3	Stack gas emission	Olafsson et al. (1989)

Source: Sigrist (1992).

Table 4.2. Examples of Selected Pollutants with Their Absorption Ranges with the CO Laser Emission Region, Their Mean Absorption Strengths, and Their Detection Limits in N_2 in Clean Air

$\lambda(\mu m)$

5.0 ⊢————————————————⊣ 6.5

Emission range of CO laser

Absorption Strength (cm^{-1}/atm)	Absorption Range of Various Gases and Vapors	Detection Limit In N_2 (ppbv)	Detection Limit In Air (ppmv)
60	Nitrogen dioxide (NO_2)	0.4	0.025
20	Acrolein (propenal) (CH_2CHCHO)	1.25	0.15
11	Acetaldehyde (CH_3CHO)	2.2	0.1
7.4	Formaldehyde ($HCHO$)	3.4	0.25
5.4	Vinyl chloride (C_2H_3Cl)	4.6	0.25
2.9	m-Xylene ($C_6H_4CH_3CH_3$)	8.6	0.5
2.2	Nitric oxide (NO)	11	0.15
2.2	1,3-Butadiene ($CH_2CHCHCH_2$)	11	0.3
1.2	Ethylene (C_2H_4)	21	1.5
1.2	Toluene ($C_6H_5CH_3$)	21	1.0
1.1	Propylene (CH_3CHCH_2)	23	1.5
0.75	Benzene (C_6H_6)	33	0.5
0.75	o-Xylene ($C_6H_4CH_3CH_3$)	33	1.0
0.55	p-Xylene ($C_6H_4CH_3CH_3$)	45	0.7
0.004	Carbon dioxide (CO_2)	6250	500
10^{-4}–2	Water vapor (H_2O)	12.5	⟨50⟩

Source: Sigrist (1992). [a] CO laser PA system [CW sealed-off CO laser (Edinburgh Instr. Type PL-3)]; ca. 90 laser lines, 1 mW–1.5 W CW.

154

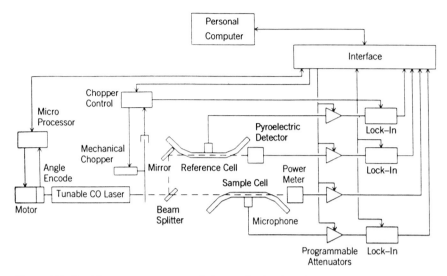

Figure 4.18. Dual-beam computer-controlled setup for CO laser PA spectroscopie gas sensing (Sigrist, 1992).

Table 4.3. CO Laser PA Analysis of Dried and Diluted Exhausts of a Volkswagon Pickup Truck Equipped with a Catalytic Converter[a]

Component	Concentration
Carbon dioxide (CO_2)	$10.8 \pm 1.1\%$
Nitric oxide (NO)	11.8 ± 1.5 ppmv
Ethylene (C_2H_4)	172 ± 10 ppmv
Propylene (C_3H_6)	109 ± 4 ppmv
Benzene (C_6H_6)	34 ± 3 ppmv
Toluene (C_7H_8)	149 ± 7 ppmv
m-Xylene (C_8H_{10})	31 ± 7 ppmv
o-Xylene (C_8H_{10})	59 ± 10 ppmv
p-Xylene (C_8H_{10})	< 1 ppmv
Formaldehyde (HCHO)	< 1 ppmv
Acetaldehyde (CH_3CHO)	3.6 ± 0.3 ppmv
Acrolein (CH_2CHCHO)	1.7 ± 0.2 ppmv

Source: Sigrist (1992).

[a] The exhausts were samples at idling operation of the cold engine.

reference cell was equipped with a movable piston at one end, which enabled the resonant frequencies of both cells to be matched. Brüel & Kjaer condenser microphones (types 4179 and 4144) were used at the cell position where the standing wave of the operating resonant mode reached the maximum amplitude. Table 4.3 shows the individual concentrations and uncertainties of a gas mixture containing 12 components. The results shown in Table 4.3 follow the analysis of data obtained using all 90 CO laser lines available. For the first time, the concentrations of 12 of the most important constituents of vehicular exhaust could be determined separately, including even different isomers of xylene. Gas chromatography can achieve separation of xylene isomers, but it lacks the ability to confirm the identity of a peak since many compounds can elute at the same retention time.

A different experimental arrangement was used with a CO_2 laser source (Sigrist, 1992). Figure 4.19 shows the setup with a low-pressure sealed-off CW $^{12}C^{16}O_2$ laser, grating-tunable to 65–70 laser transitions between 9.2 and 10.8 µm. The PA cell was operated at a radial resonance frequency at 2650 Hz with a Q value of 340 for pure N_2 at 296 K (Meyer and Sigrist, 1990). Table 4.4 shows the list of gases investigated with the apparatus of Fig. 4.19. The cell was characterized by a minimum detectable absorption coefficient of $1–3 \times 10^{-8}$ cm^{-1}, depending on the wavelength, for a laser power of 1 W.

Several in situ air pollution monitoring studies of vehicular exhausts were performed with this apparatus (e.g. as shown in Fig. 4.20), as well as industrial pollutant sensing (Fig. 4.21). Table 4.5 shows the results of the PA analysis

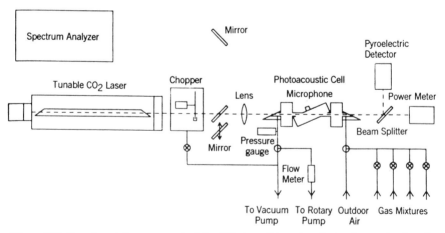

Figure 4.19. Experimental arrangement of the CO_2 laser PA system for in situ monitoring of trace gases (Sigrist, 1992).

Table 4.4. List of Gases Investigated with the CO_2 Laser Photoacoustic System of Fig. 4.19

Substance	Formula	Mol. Weight (g/mol)	Sat. Vapor Press. at $T = 20\,°C$ (mbar)	Conv. Factor at $T = 20\,°C$ ($\mu g \cdot m^{-3} \cdot ppbv^{-1}$)	σ^{max} at $^{12}C^{16}O_2$ Laser Transitions 10^{-20} cm^2	Transition[a]
Benzene	C_6H_6	78.11	99.4	3.27	7.2	9 P(30)
1,3-Butadiene	C_4H_6	54.09	>1000	2.26	2.8	10 P(46)
Carbon dioxide	CO_2	44.01	>1000	1.84	0.015	10 R(16)
Chlorobenzene	C_6H_5Cl	112.56	11.7	4.71	13	9 R(32)
o-Dichlorobenzene	$C_6H_4Cl_2$	147.01	1.3	6.15	21	9 P(28)
m-Dichlorobenzene	CH_4Cl_2	147.01	2.0	6.15	11	9 R(28)
p-Dichlorobenzene	CH_4Cl_2	147.01	1.7	6.15	5.5	9 R(34)
Dichloromethane	CH_2Cl_2	84.93	472.7	3.54	0.17	10 P(44)
Ethanol	C_2H_5OH	46.07	58.5	1.93	17	9 R(16)
Ethylbenzene	C_8H_{10}	106.16	9.42	4.44	2.3	9 P(36)
Ethylene	C_2H_4	28.05	>1000	1.17	130	10 P(14)
Methanol	CH_3OH	32.04	129.3	1.34	72	9 P(34)
Propylene	C_3H_6	42.08	>1000	1.76	8.2	10 P(36)
Toluene	C_7H_8	92.13	29.1	3.85	4.5	9 P(36)
Vinylchloride	C_2H_3Cl	62.50	>1000	2.61	29	10 P(22)
Water	H_2O	18.02	23.3	0.75	0.005	10 P(40)
o-Xylene	C_8H_{10}	106.16	6.49	4.44	4.3	9 P(12)
m-Xylene	C_8H_{10}	106.16	8.17	4.44	2.8	9 P(24)
p-Xylene	C_8H_{10}	106.16	8.65	4.44	2.7	9 P(38)

Source: Sigrist (1992).

[a] P transitions occur from the $J \rightarrow J + 1$ rotational level of the CO_2 laser. R transitions occur from the $J \rightarrow J - 1$ rotational level.

157

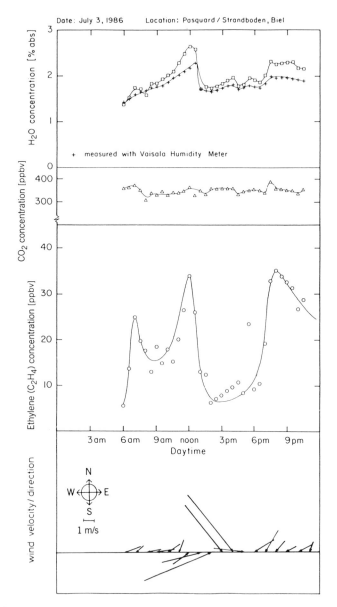

Figure 4.20. Temporal H_2O vapor, CO_2, and C_2H_4 concentration profiles, derived from PA measurements and wind data obtained by commercial equipment on location at a city park in Biel, Switzerland, on July 3, 1986 (Sigrist, 1992).

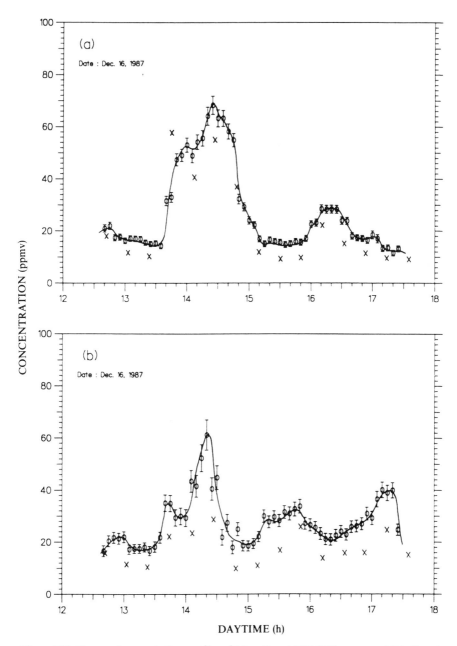

Figure 4.21. Temporal concentration profiles of (a) methanol (CH_3OH)–vapor and (b) ethanol (C_2H_5OH)–vapor of emission source A of a pharmaceutical production plant in Basel, Switzerland. The concentrations were derived from PA measurements (○) and compared to gas chromatographic (GC) data (×) (Sigrist, 1992).

Table 4.5. CO$_2$ Laser PA Analysis of Industrial Exhausts (Emission Sources B)

Component	Concentration
Water (H$_2$O)	$1.7 \pm 0.3\%$ abs.[a]
Carbon Dioxide (CO$_2$)	466 ± 355 ppmv
Toluene (C$_7$H$_8$)	47.5 ± 13.6 ppmv
Chlorobenzene (C$_6$H$_5$Cl)	27.8 ± 3.1 ppmv
m-Dichlorobenzene (C$_6$H$_4$Cl$_2$)	80.8 ± 3.2 ppmv
o-Dichlorobenzene (C$_6$H$_4$Cl$_2$)	4.4 ± 1.2 ppmv
Ethanol (C$_2$H$_5$OH)	6.4 ± 1.8 ppmv
Methanol (CH$_3$OH)	0.2 ± 0.4 ppmv
Ethylene (C$_2$H$_4$)	1.3 ± 0.4 ppmv

Source: Sigrist (1992).
[a] In comparison, a capacitive humidity meter yielded 1.5% for the absolute H$_2$O vapor concentration.

from an industrial emission site. Sigrist (1992) suggested a combination of PA spectroscopy and gas chromatography to improve the detection selectivity in cases of dominant absorption by a single substance in a mixture of gases (Nickolaisen and Bialkowski, 1986). The use of PA spectroscopy as a gas-sensing technique is fast maturing in the direction of multicomponent trace gas detection, yet the complexity and expense of the current PA systems required for this task may deter many potential users from employing this technique's superior sensitivity and opt for, say, semiconductor sensor arrays (see Chapter 3).

REFERENCES

Adler-Golden, S.M., Goldstein, N., and Bien, F. (1990). U.S. Patent 4, 953, 976.

Adler-Golden, S.M., Goldstein, N., Bien, F., Matthew, M.W., Gersh, M.E., Cheng, W.K., and Adams, F.W. (1992). *Appl. Opt.* **31**, 831.

Aspnes, D.E., and Studna, A.A. (1975). *Appl. Opt.* **14**, 220.

Assam, R.M.A., and Bashara, N.M. (1977). *Ellipsometry and Polarized Light.* North-Holland Publ., New York.

Bernegger, S., and Sigrist, M.W. (1987). *Appl. Phys.* **B44**, 125.

Bernegger, S., and Sigrist, M.W. (1990). *Infrared Phys.* **30**, 375.

Bernegger, S., Meyer, P.L., Wildmer, C., and Sigrist, M.W. (1988). *Springer Ser. Opt. Sci.* **58**, 122.

Bialkowski, S.E., and Long, G.R. (1987). *Anal. Chem.* **59**, 873.

Bicanič, D., Harren, F., Reuss, J., Woltering, E., Snel, J., Voesenck, L.A.C.J., Zuidberg, B., Jalinek, H., Bijnen, F., Blom, C.W.P.M., Sauren, H., Kooijman, M., van Hove, L., and Tonk, W. (1989). *Springer Ser. Top. Curr. Phys.* **46**, 213.

Burkhardt, E.G., Lambert, C.A., and Patel, C.K.N. (1975). *Science* **188**, 1111.

Fortunato, G., D'Amico, A., Coluzza, C., Sette, F., Capasso, C., Patella, F., Quaresima, C., and Perfetti, P. (1984). *Appl. Phys. Lett.* **44**, 887.

Fortunato, G., Bearzotti, A., Caliendo, C., and D'Amico, A. (1989). *Sens. Actuators* **16**, 43.

Frans, S.D., and Harris, J.M. (1985). *Anal. Chem.* **57**, 2680.

Frazier, G.A., and Glosser, R. (1982). *Solid State Commun.* **41**, 245.

Goldner, R.B., Mendelsohn, D.H., Alexander, J., Henderson, W.R., Fitzpatrick, D., Haas, T.E., Sample, H.H., Rauh, R.D., Parker, M.A., and Rose, T.L. (1983). *Appl. Phys. Lett.* **43**, 1093.

Haaland, D.M. (1990). In *Practical Fourier Transform Infrared Spectroscopy* (J.R. Ferraro and K. Krishnan, Eds.). Chapter 3. Academic Press, San Diego.

Hess. P. (1983). *Top. Curr. Chem.* **111**, 1.

Hess, P. (1992). In *Principles and Perspectives of Photothermal and Photoacoustic Phenomena* (A. Mandelis, Ed.), Chapter 4. Elsevier, New York.

Ito, K., and Ohgami, T. (1992). *Appl. Phys. Lett.* **60**, 938.

Kerr, E.L., and Atwood, J.G. (1968). *Appl. Opt.* **7**, 915.

Kohn, H.W., and Boudart, M. (1964). *Science* **145**, 149.

Kretschmann, E., and Raether, H. (1968). *Z. Naturforsch. A* **23**, 2135.

Kreuzer, L.B. (1971). *J. Appl. Phys.* **42**, 2934.

Kreuzer, L.B. (1977). In *Optoacoustic Spectroscopy and Detection* (Yoh-Hon Pao, Ed.), Chapter 1. Academic Press, New York.

Kreuzer, L.B., and Patel, C.K.N. (1971). *Science* **173**, 45.

Liedberg, B., Nylander, C., and Lundström, I. (1983). *Sens. Actuators* **4**, 299.

Lloyd, J.P., Pearson, C., and Petty, M.C. (1988). *Thin Solid Films* **160**, 431.

Lundström, I. (1981). *Sens. Actuators* **1**, 403.

Maeda, M., Takahashi, N., and Kuwano, Y. (1990). *Sens. Actuators* **B1**, 215.

Margaritondo, G. (1983). *Solid-State Electron.* **26**, 499.

Martens, H., and Nase, T. (1989). *Multivariate Calibration*. Wiley, Chichester, UK.

Martensson, J., Arwin, H., and Lundström, I. (1990). *Sens. Actuators* **B1**, 134.

Meyer, P.L., and Sigrist, M.W. (1990). *Rev. Sci. Instrum.* **61**, 1779.

Meyer, P.L., Bernegger, and Sigrist, M.W. (1988). *Springer Ser. Opt. Sci.* **58**, 122.

Nickolaisen, S.L., and Bialkowski, S.E. (1986). *J. Chromatogr.* **366**, 127.

Nylander, C., Liedberg, B., and Lind, T. (1982). *Sens. Actuators* **3**, 79.

Olafsson, A., Hammerich, M., Bülow, J., and Henningsen, J. (1989). *Appl. Phys.* **B49**, 91.

Patel, C.K.N. (1976). *Opt. Quantum Electron.* **8**, 145.

Patel, C.K.N., Burkhardt, E.G., and Lambert, C.A. (1974). *Science* **184**, 1173.

Perlmutter, P., Shtrikman, S., and Slatkine, M. (1979). *Appl. Opt.* **18**, 2267.

Petty, M.C. (1987). In *Polymer Surfaces and Interfaces* (W.J. Feast and H.S. Munro, Eds.), p. 163. Wiley, New York.

Raether, H. (1977). *Phys. Thin Films* **9**, 145.

Reihua, W., Fortunato, G., and D'Amico, A. (1985). *Sens. Actuators* **7**, 253.

Rooth, R.A., Verhage, A.J.L., and Wouters, L.W. (1990). *Appl. Opt.* **29**, 3643.

Sauren, H., Bicanič, D., and Jalink, H. (1989). *J. Appl. Phys.* **66**, 5085.

Schoch, K.F., and Temofonte, T.A. (1988). *Thin Solid Films* **165**, 83.

Sigrist, M.W. (1992). In *Principles and Perspectives of Photothermal and Photoacoustic Phenomena* (A. Mandelis, Ed.), Chapter 7. Elsevier, New York.

Sigrist, M.W., Bernegger, S., and Meyer, P.L. (1989a). *Springer. Ser. Top. Curr. Phys.* **46**, 173.

Sigrist, M.W., Bernegger, S., Meyer, P.L. (1989b). *Infrared Phys.* **29**, 805.

Tam, A.C. (1983). In *Ultrasensitive Laser Spectroscopy* (D.S. Kliger, Ed.), Chapter 1. Academic Press, New York.

Weber, J.H., and Busch, D.H. (1965). *Inorg. Chem.* **4**, 469.

Wicke, E., and Brodowsky, H. (1978). *Springer Top. Appl. Phys.* **29**, 73.

Wolff, E.W., and Peel, D.A. (1985). *Nature* (*London*) **313**, 535.

Zhu, D.G., Petty, M.C., and Harris, M. (1990). *Sens. Actuators* **B6**, 265.

CHAPTER

5

FIBER-OPTIC SENSORS

5.1. INTRODUCTION AND HISTORICAL PERSPECTIVE

The concept of fiber-optic sensors (FOS) is far from new. The first patents were obtained around the year 1965 (see Dakin and Culshaw, 1988). In the same year Gamble et al. (1965) reported the use of fiber optics for a clinical cardiac characterization. However, only after considerable experience accumulated with conventional gas analyzer spectrometers did serious research efforts toward development of fiber-optic detector technology begin in the early 1980s (see Kotte et al., 1989).

An optical fiber is essentially a "light pipe," or waveguide, for optical frequencies. It is typically drawn from a spherical mirror to a diameter of a few to a few hundred micrometers. A large variety of optical fibers is available from commercial manufacturers, and these can be divided into five main types: all-polymer, silica, plastic-coated silica, glass fibers, and fiber bundles (Dakin and Culshaw, 1988). One of the most valuable properties of an optical fiber is its flexibility, because of which it can be used for the transfer of optical signals over distances of kilometers without the necessity of perfect alignment between source and detector and especially without much attenuation or signal degradation due to ambient electrical noise or electromagnetic interferences (Miller et al., 1973; Giallorenzi, 1978). This property, among others, has led to many applications in the area of optical communication and in areas such as telephony and data transmission.

Recently there has been considerable research activity in the development of FOS in the field of gas sensing. The measurement basis in FOS consists of changing the features of light transmitted along the fiber, and these changes in turn are used to modify an output electrical signal in a receiver. FOS devices provide a means whereby light guided within an optical fiber can be modified in response to an external physical, chemical, biological, or other such influence. Research efforts toward the development of gas fiber-optic detectors have progressed at a quick pace (Janata, 1989). Optical fibers acting as light carriers bring several attractive features to the chemical sensor field. In 1975, Hardy et al. reported the first FOS device for chemical gas detection. They used a FOS for the detection of cyanide (CN^-) traces. A year later,

163

David et al. (1976) showed the ability of a FOS device to detect ammonia in ambient air. Giuliani et al. (1983, 1985) have also detected ammonia vapor with a FOS device. FOS have further been used for the detection of oxygen, glucose, carbon, CH_4, carbon dioxide, and various other gases (Narayanaswamy, 1987; Harmer and Narayanaswamy, 1988). Several review papers concerning gas fiber optics describe in detail the development of many FOS for the detection of several gases (Nylander, 1985; Seitz, 1984; Chabay, 1982; Harmer, 1987; Norris, 1991). The capability of FOS for the detection of oxygen and carbon dioxide has been exploited by Wolfbeis et al. (1988). Butler (1984) was the first to demonstrate that an FOS device with appropriate chemically sensitive coating (Pd) could be employed as a hydrogen detector. In this chapter, we review the development of fiber-optic gas detectors and mainly the work of Butler (1984, 1990) and Butler and Ginley (1987, 1988a), who developed the capability of the Pd–FOS device as a hydrogen detector.

5.2. THEORETICAL BACKGROUND

We can distinguish two main kinds of FOS (Seitz, 1984; Chabay, 1982; Harmer, 1987; Butler, 1984): (a) *extrinsic*—the light can be allowed to exit from the optical fiber and to interact with various media before continuing to propagate inside the same or another fiber; and (b) *intrinsic*—the light continues to propagate in the optical fiber without any outside interaction (see Fig. 5.1).

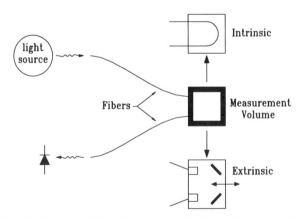

Figure 5.1. Schematic diagram of intrinsic and extrinsic fiber-optic sensors (Christofides and Mandelis, 1990).

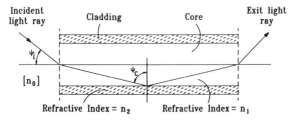

Figure 5.2. Light propagation through an optical fiber (Harmer and Narayanaswamy, 1988).

The principle of the FOS for gas detection is straightforward. When gas molecules are absorbed in the active (catalyst-coated) fiber part, which is chosen according to the desired gas selectivity, they change the optical properties of the coated layer. Depending on the particular device, the optical property measured can be absorptance, reflectance, luminescence, or scattering. The propagation of the light in an optical fiber can be explained with various degrees of accuracy in terms of electromagnetic theory or geometric optics.

Figure 5.2 shows the principle of light propagation in an optical fiber with the refractive indices n_1 and n_2 for core and cladding, respectively. In the case of the optical fiber, $n_1 > n_2$ in order to promote total internal reflection. An incident ray that reaches the fiber at an angle θ_i is reflected at the critical angle θ_c from the core/cladding interface, where θ_c is given by

$$\sin \theta_c = \frac{n_2}{n_1} \qquad (5.1)$$

From geometric optics (Snell's law), it is easy to show that

$$n_0 \sin \theta_i = (n_1^2 - n_2^2)^{1/2} \qquad (5.2)$$

where n_0 is the refractive index of the medium external to the optical fiber ($n_0 = 1$ when the external medium is air). The right-hand side of Eq. (5.2) is known as the numerical aperture (NA) of the fiber. If the outside medium is air, then Eq. (5.2) can be rewritten as

$$NA = \sin \theta_i \qquad (5.3)$$

It is obvious from Fig. 5.2 that the light propagation in the optical fiber is dependent only upon the refractive indices of the core and cladding. A large difference in the values of these indices is necessary for a large acceptance angle (Dakin and Culshaw, 1988). For a typical fiber, $\theta_i \approx 10°$.

5.3. FIBER-OPTIC INSTRUMENTATION

Interferometry is the most useful measurement mode applied to FOS technology. The physics of dual-beam interference is very old. However, the concept of this phenomenon is extremely useful in practice in the field of fiber-optic detectors. It is well known that the maxima of the resulting fringe pattern appear where the phase difference between the interfering beams is $2K_n\pi$ (K_n is a positive integer). Any small perturbation in the phase of one of the beams due to external changes (temperature, pressure, magnetic field, chemical gas, etc.) will cause a transverse shift in position of the fringe pattern. According to Dakin and Culshaw (1988) the resolution of modern optoelectronic techniques is about 10^{-4} of the fringe spacing, whereas according to Harmer (1987) this resolution is even better and close to 10^{-6}. Many kinds of interferometer have been developed for gas detection; in Fig. 5.3 we present the popular Mach–Zehnder interferometer as representative of FOS technology. In recent years special spectrometers have been developed for fiber optics in order to adapt them to FOS technology (Butler, 1984; Butler and Ginley, 1988a, b). Conventional spectrometers have been modified to allow miniaturization, integration with the processing electronics, and multiple simultaneous connection with many fibers. According of Laude et al. (1983) a small modern spectrometer equipped with an optical fiber can be constructed from a small block of BK 7 glass ($20 \times 20 \times 98 \text{ mm}^3$) with a spherical mirror on one end and a grating on the other. An example of miniaturization of a FOS on a printed circuit card inside the electronics has been reported by Körth (1983, 1984) and Tien and Capik (1980). In Fig. 5.4 in Section 5.5.1 (below) a new version of the Mach–Zehnder interferometer will be shown that uses optical fibers instead of mirrors (Butler, 1984; Butler and Ginley, 1988a, b). This version has been used for the development of FOS in the field, with the advantage of detector portability.

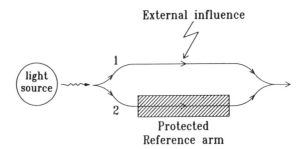

Figure 5.3. Optical fiber Mach–Zehnder interferometer (Butler, 1984).

5.4. MEASUREMENT TECHNIQUES

Fiber-optic sensors for gas detection exhibit a large number of measurable parameters such as refractivity, reflectivity, absorption, and scattering. In essence there are three main modes of detection for monitoring gases with FOS (Harmer and Narayanaswamy, 1988), as discussed in the following three subsections.

5.4.1. Absorptance Measurements

In this case the light intensity shift is determined by the quantity of absorbing species (gas molecules or atoms) in the optical path and is related to the Beer–Lambert relationship:

$$A_I = \log \frac{I_0}{I} = m_a l C_0 \qquad (5.4)$$

where A_I is the absorptance; I_0 and I are the incident and the transmitted light intensity, respectively; l is the path length of the light; C_0 is the concentration of the absorbing species; and m_a is the gas molar absorptivity. In the case where the medium and/or the selective chemistry used in a FOS does not allow any transmission of light, a measurement of the intensity of the reflected light may be used.

5.4.2. Reflectance Measurement

The optical characteristics of diffuse reflectance are a function of the composition of an optically excited system. According to the Kubelka–Munk theory (Kubelka and Munk, 1931), the reflectance R_∞ of a semi-infinite optical medium is related to the absorption coefficient α and the scattering coefficient S_c (which is assumed to be independent of the concentration) through the relation

$$f(R_\infty) = \frac{(1 - R_\infty)^2}{2R_\infty} = \frac{\alpha}{S_c} = m_a C_0 \qquad (5.5)$$

where $f(R_\infty)$ is known as the Kubelka–Munk function and has the aforementioned simple dependence on R_∞ in a semi-infinite sample.

5.4.3.   Luminescence (or Fluorescence) Measurements

This kind of monitoring is very useful for detection of very low concentrations. The intensity of fluorescence I_F is given by the relationship (Harmer and Narayanaswamy, 1988):

$$I_F = E_x I_0 \Phi_F m_a l C_0 \qquad (5.6)$$

where E_x is an experimental constant related to the instrumentation and sensor configuration, and Φ_F is the quantum yield of fluorescence. The rest of the symbols have their earlier definitions. The linearity of Eq. (5.6) as a function of C_0, however, is violated at high concentrations.

5.5.   FIBER-OPTIC GAS DETECTION

5.5.1.   Hydrogen Detection

In this subsection we consider experimental results obtained by Butler (1984) concerning the use of a Pd-coated fiber-optic detector (Pd–FOS) for hydrogen detection. A Mach–Zehnder interferometer was used. The experimental setup is shown in Fig. 5.4(a). Both ends of the coated and uncoated fibers were glued to a fused quartz plate with Eastman 910 adhesive. A more detailed description of the apparatus has been given by Butler and Ginley (1988b) [see Fig. 5.4(b)]. The Pd coating was made with the sputtering method, and the thickness was 1.5 μm. Owing to the close match in index of refraction between the coated and uncoated fibers, this arrangement makes an effective mode stripper. The light of a 0.5-mW He–Ne laser is split into two directions: one through the Pd-coated fiber, and the other through the uncoated fiber. Both are exposed to the gas flow (fiber diameter: 80 μm). As shown in Fig. 5.5, the movement of the fringe pattern can be observed visually or by use of a simple photodetector and chart recorder. Figure 5.5 presents the Pd–FOS response as a function of time for 0.6% of H_2 in N_2. The movement of the fringe pattern by one fringe corresponds to a change in optical path length of the arm of the interferometer by one wavelength of light. According to Butler (1984) the introduction of hydrogen (0.6%) in the test cell volume of approximately 75 ml gas causes the passage of three fringes (therefore one fringe corresponds to 0.2% H_2 in N_2). Butler and Ginley (1988a) performed some further experiments using a hydrogen-sensitive palladium-coated optical fiber in the range of 20 ppb to 10^4 ppm under STP conditions in nitrogen gas, which represents a record of sensitivity to H_2 gas under flow through and STP conditions. Figure 5.6 presents the response of the Pd–FOS detector

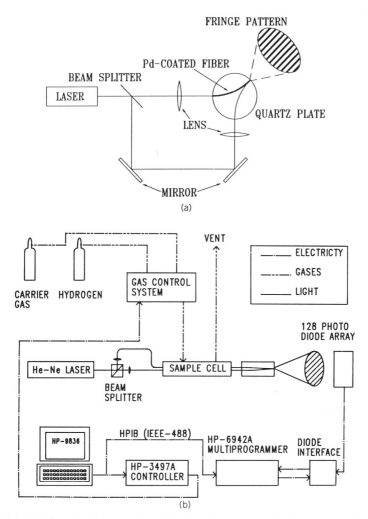

Figure 5.4. (a) Schematic of the Mach–Zehnder interferometer used for hydrogen gas detection by Butler (1984). (b) Complete FOS apparatus for hydrogen gas detection. HP, Hewlett–Packard; HPIB, Hewlett–Packard Interface Bus. (Butler and Ginley, 1988).

in a wide range of hydrogen concentration on N_2. In this figure note that the authors reported results down to 1 ppb, although the lower limit of sensitivity was estimated to be a few ppb. This estimation seems to be realistic owing to the fact that early measurement capabilities achieved lower resolution than did modern optoelectronic techniques (Butler and Ginley, 1988a, b): the latter can measure better than 10^{-4} of the fringe spacing. This phenomenon

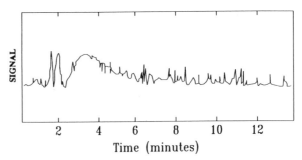

Time (minutes)

Figure 5.5. Pd–FOS response as a function of time, for 0.6% of hydrogen in nitrogen (flow rate: 400 mL/min) (Butler, 1984).

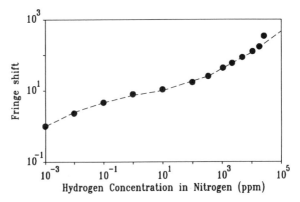

Hydrogen Concentration in Nitrogen (ppm)

Figure 5.6. The fringe shift for the Pd–FOS detector vs. hydrogen concentration (Butler and Ginley, 1988a).

is reversible, and a similar behavior is observed when H_2 is removed from the Pd film. The temperature fluctuation effects were found to be equivalent to 2 ppm H_2/°C. This noise level could potentially limit the excellent resolution of the FOS. Thus the position of the fringes as a function of time gives much information on the kinetics of interaction of the Pd–H_2 system.

In order to interpret their experimental results, Butler (1984) and Butler and Ginley (1988a, b) limited their analyses to the pressure regime in which only the α-phase of Pd exists. Then, by using a theoretical relationship between the hydrogen partial pressure, $p(H_2)$, and the hydride composition in the Pd, they expressed the axial and radial strains as functions of Pd coating thickness in units of optical fiber radius. These investigators used the Hughes and Jarzynski (1980) relation

$$\frac{\Delta\phi}{\phi} = w - \frac{n^2}{2}[2u(P_{11} - P_{44}) + w(P_{11} - 2P_{44})] \qquad (5.7)$$

where $n = 1.46$ is the index of refraction at the fiber center; P_{11} and P_{44} are Pockel's coefficients and are equal to 0.1254 and 0.0718, respectively; and u and w depend on the Pd thickness (Butler, 1984; Butler and Ginley, 1988a, b). By using the above equation one can calculate the phase shift $\Delta\phi$ due to the change in optical path length. However, this semiquantitative analysis led to a 60% discrepancy between theoretical and experimental results. Butler (1984) justified this discrepancy as being a result of the variation of palladium properties, nonuniformity of the Pd thickness, etc. However, it ought to be remembered that this discrepancy may arise because for hydrogen concentration around 0.6% the β-phase transition becomes significant.

In 1991, Butler reported the first hydrogen FOS able to provide a large signal for concentrations of hydrogen in air near the explosive limit. Figure 5.7 shows the basic configuration of the active element. A detailed description of a class of FOS based on a chemically sensitive layer deposited on the cladding of an optical fiber has been given earlier by Butler and his co-workers (see Butler, 1990; Butler et al., 1990a, b, Butler and Ricco, 1988, 1989). A typical response of the FOS presented in Fig. 5.7 is given in Fig. 5.8. We

Optical Fiber

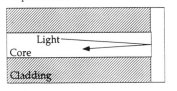

PdH$_x$ Film

Figure 5.7. Representation of the micromirror hydrogen sensor. The fiber is 125 μm in diameter with a 50 μm core. The palladium film is deposited by evaporation and is typically 10–20 nm thick (Butler, 1991).

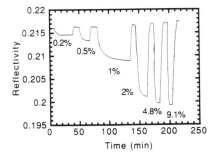

Figure 5.8. Response of the 10 nm thick Pd micromirror hydrogen FOS varying concentrations of hydrogen in nitrogen. Between exposures to hydrogen, the sensor is exposed to air. The measurements were performed at room temperature (Butler, 1991).

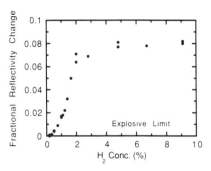

Figure 5.9. Magnitude of the response of the micromirror hydrogen sensor at room temperature to varying concentrations of hydrogen in air. Note that the hydrogen actually produces a decrease in reflectivity as shown in Fig. 5.7. The absolute reflectivity of this film is about 22%. (Butler, 1991).

note the change in reflectivity as a function of time for various hydrogen concentrations. It has been noted by Butler (1991) that the magnitude of the change in reflectivity, upon exposure to a given $[H]$, increases monotonically with Pd film thickness. Butler (1991) has also reported that while the behavior of such FOS depends on the Pd deposition condition, several exposures to hydrogen (about 20 cycles) suffice for the stabilization of the device. This phenomenon has also been described by Christofides and Mandelis (1989) for their Pd–PPE device.

Figure 5.9 shows the response of the FOS to various concentrations of hydrogen in air. The sensor has been found to be completely reversible and reproducible. A rapid increase of the FOS response around the 2% of $[H]$ can be observed. All these measurements have been performed at STP conditions up to 100% hydrogen with no damage to the sensor or any changes in the reproducibility of the device. From 10 to 100% H_2 in air the reflectivity continues to increase slowly. This type of response arises because of the complex nature of PdH_x system. The large change of reflectivity that occurs around 2% in Fig. 5.9 is due to the Pd phase transformation (Lewis, 1967), while the slower increase in signal above the 2% is due to the slower compositional change of β-phase Pd hydride with hydrogen partial pressure. The critical concentration for phase transformation is strongly dependent on the working temperature (Lewis, 1967).

5.5.2. Chemisorption Measurements

Purely gas detection is not the only application of FOS. In fact, during the last few years FOS has been extensively used for the study of chemisorption of molecules on metal surfaces (Butler and Ricco, 1988; Butler et al., 1990a, b). Butler and his co-workers demonstrated that the reflectivity of an optically thin metal film at the tip of a single optical fiber is sensitive to the chemisorption of monolayers of small molecules such as SO_2, H_2S, CO, and CO_2.

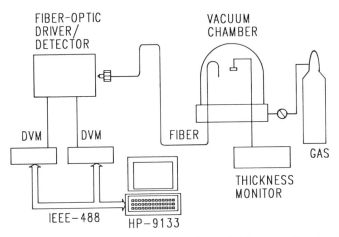

Figure 5.10. Apparatus used to measure the reflectivity of thin metal films DVM, digital voltmeter; HP, Hewlett–Packard. (Butler, 1990)

Figure 5.10 shows the apparatus that has been used to measure the reflectivity of thin metal films. In 1990, Butler et al. (1990a, b) described a similar optical technique for studying the interaction between Hg vapor and a thin Au film surface.

5.6. COMPARISON BETWEEN PURELY ELECTRONIC AND FIBER-OPTIC SENSORS

It appears that coherent light does seem to be very competitive with modern electronics as a hydrogen-sensing principle. The propagation of light waves is, however, very sensitive to the absorption, reflection, and transmission in the propagation medium. Nevertheless, the choice of wavelength-selective fibers has minimized—and even nullified—these problems. The material for the optical waveguide is carefully chosen for selected wavelengths at which dispersion and attenuation in the fiber are minimal. In what follows we describe some of the attractive features and disadvantages of chemical sensors based on fiber optics: (a) The optical nature of the signal does not introduce any electrical interference. This advantage makes FOS very useful in environments with considerable electromagnetic activity. (b) The most exciting possibility offered by chemical fiber optics appears to be (Seitz, 1984) the use of multiwavelength and time-domain information. For example, one envisages sensors that report simultaneously the detection of two or more pollutants. Naturally, it is important to note that order detectors described in previous sections also share this advantage. However, for simultaneous detection of different gases the coated catalyst has an important and well-defined role to play. (c) Optical sensors can be developed to respond in environments incompatible with electroded devices. (d) Their small size makes optical fibers attractive for beam guidance to and from remote sensors.

Unfortunately, FOS are subject to several limitations as compared to electronic solid state devices. The most important is that ambient light could interfere with the background noise level. Thus FOS require a dark environment for optimal operations and noise minimization.

Table 5.1. More General Applications of FOS Devices

References	FOS
Bucaro et al. (1977)	Acoustic
Bucaro and Hickman (1979)	Acoustic
Bucaro et al. (1977)	Pressure
Cole et al. (1977)	Pressure
Culshaw et al. (1977)	Pressure
Hocker (1979)	Pressure
Othonos et al. (1993)	Pressure
Hocker (1979)	Temperature
Heaton (1980)	Temperature
Lagakos et al. (1981)	Temperature
Dils et al. (1986)	Temperature
Othonos et al. (1993)	Temperature
Leslie et al. (1981)	Spectrophone
Hocker (1979)	Hydrophone
Hughes and Jarzynski (1980)	Hydrophone
Yariv and Winsor (1980)	Magnetic
Jarzynski et al. (1980)	Magnetic
Peterson et al. (1980)	pH monitor
Smela and Santiago-Aviles (1988)	Detection in liquids
Butler and Ricco (1988)	Reflectance
Emge and Chen (1991)	2D tactile imager[a]
Szarka (1988)	Military
Kawaguchi et al. (1991)	Biology
Kraus et al. (1988)	Photometry

[a]2D = two-dimensional.

5.7. OTHER APPLICATIONS OF FIBER-OPTIC SENSORS

Detection of pressure with an optical fiber positioned in the arms of a Mach–Zehnder interferometer was reported by Hocker (1979). The measurement of pressure by means of optical fibers is effected by the variations in the relative phase between light propagating in the two interferometer arms. This variation of phase is attributed to the changes in the optical propagation characteristics in the arm exposed to the applied pressures (Hocker, 1979; Hughes and Jarzynski, 1980). On the other hand, the change in the optical fiber length due to thermal expansion or contraction, as well as the change induced by temperature in its refraction characteristics, led to the development of the optical fiber temperature detector (Hocker, 1979). In 1981, Leslie et al. reported the first fiber-optic spectrophotometer in which the pressure-sensing transducer was contacted using a fiber-optic acoustic-sensing coil in one of the arms of an interferometric fiber sensor. A fiber-optic hydrophone has also been reported (Hocker, 1979; Hughes and Jarzynski, 1980). Yariv and Winsor (1980) demonstrated theoretically the possibility of detecting weak magnetic fields by using an optical fiber with a magnetostrictive jacket, and Jarzynski et al. (1980) followed up with experimental verification. Seiler and Leach (1982) studied the thermally induced phase noise in a fiber-optic interferometer and particularly in the magnetic field FOS. A fiber-optic probe for pH monitoring was also reported by Peterson et al. (1980). An optical fiber sensor able to detect water and solvent in oil was reported by Smela and Santiago-Aviles (1988). FOS for biomedical applications have been developed by Peterson and Vurek (1984). In Table 5.1 are listed the different types of FOS that have been developed during the last few years for the measurement of various physical components.

REFERENCES

Bucaro, J.A., and Hickman, T.R. (1979). *Appl. Opt.* **18**, 938.

Bucaro, J.A., Dardy, H.D., and Carome, E.F. (1977). *J. Acoust. Soc. Am.* **62**, 1302.

Butler, M.A. (1984). *Appl. Phys. Lett.* **45**, 1007.

Butler, M.A. (1990). *Proc. SPIE: Chem. Biochem. Environ. Fiber Sens. II* **1368**, 46.

Butler, M.A. (1991). *J. Electrochem. Soc.* **138**, L46.

Butler, M.A., and Ginley, D.S. (1987). *Proc.—Electrochem. Soc.* **87-9**, 502.

Butler, M.A., and Ginley, D.S. (1988a). *J. Appl. Phys.* **64**, 3706.

Butler, M.A., and Ginley, D.S. (1988b). *J. Electrochem. Soc.* **135**, 45.

Butler, M.A., and Ricco, A.J. (1988). *Appl. Phys. Lett.* **53**, 1471.

Butler, M.A., and Ricco, A.J. (1989). *Sens. Actuators* **19**, 249.

Butler, M.A., Ricco, A.J., and Baughman, R.J. (1990a). *J. Appl. Phys.* **67**, 4320.

Butler, M.A., Ricco, A.J., and Buss, R. (1990b). *J. Electrochem. Soc.* **137**, 1325.

Chabay, I. (1982). *Opt Waveguides Anal. Chem.* **54**, 1071A.

Christofides, C., and Mandelis, A. (1989). *J. Appl. Phys.* **66**, 3975.

Christofides, C., and Mandelis, A. (1990). *J. Appl. Phys.* **68**, R1.

Cole, J.H., Johnson, R.L., and Buta, P.G. (1977). *J. Acoust. Soc. Am.* **62**, 1136.

Culshaw, B., Davis, D.E.N., and Kingsley, S.A. (1977). *Electron. Lett.* **13**, 760.

Dakin, J., and Culshaw, B. (1988). *Optical Fiber Sensors: Principles and Components.* Artech House, Boston.

David, D.J., Wilson, M.C., and Ruffin, D.S. (1976). *Anal. Lett.* **9**, 389.

Dils, R.R., Geist, J., and Reilly, M.L. (1986). *J. Appl. Phys.* **59**, 1005.

Emge, S.R., and Chen, C.L. (1991). *Sens. Acuators* **B3**, 31.

Gamble, B.G., Hugenholtz, P.G., Monroe, R.G., Polanyi, M., and Nadas, A.S. (1965). *Intracardiac Oximetry Circ.* **31**, 328.

Giallorenzi, T.G. (1978). *Proc. IEEE* **66**, 744.

Giuliani, J.F., Wohltjen, H., and Jarvis, N.L. (1983). *Opt. Lett.* **8**, 54.

Giuliani, J.F., Bey, P.P., Jr., Wohltjen, H., Snow A., and Jarvis, N.L. (1985). *IEEE Int. Conf. Solid-State Sens. Actuators*, New York, p. 74.

Hardy, E.E., David, D.J., Kapany, N.S., and Unfortunately, F.C. (1975). *Nature (London)* **257**, 666.

Harmer, A.L. (1987). *Proc. Symp. Chem. Sensors, Electrochem. Soc.* **87-9**, 409.

Harmer, A.L., and Narayanaswamy, R. (1988). In *Chemical Sensors* (T.E. Edmonds, Ed.), Chapter 13. Chapman & Hall, New York.

Hauden, D., Jaillet, G., and Coquerel, R. (1981). *Proc. IEEE Ultrason. Symp., 1981* p. 148.

Heaton, H. (1980). *Appl. Opt.* **19**, 3719.

Hocker, G.B. (1979). *Appl. Opt.* **18**, 1445.

Hughes, R., and Jarzynski, J. (1980). *Appl. Opt.* **19**, 98.

Janata, J. (1989). *Principles of Chemical Sensors*, Chapter 5. Plenum, New York.

Jarzynski, J., Cole, J.H., Bucaro, J.A., and Davis, C.M. (1980). *Appl. Opt.* **19**, 3746.

Kawaguchi, T., Shiro, T., and Iwata, K. (1991). *Sens. Actuators.* **B3**, 113.

Körth, H.E. (1983). *J. Phys. (Paris)* **44** (C10), 101.

Körth, H.E. (1984). *Int. Conf. Opt. Fiber Sens. 2nd*, 219.

Kotte, E., Derge, K., Landeryou, R.R., Propawe, P., Tschudi, T., and Wobbe, W. (1989) *Technologies of Light*, Chapter 4. Springer-Verlag, Berlin.

Kraus, P.R., Wade, A.P., Crouch, S.R., Holland, J.F., and Miller, B.M. (1988). *Anal. Chem.* **60**, 1387.

Kubelka, P., and Munk, F. (1931). *Z. Tech. Phys.* **12**, 593.

Lagakos, N., Bucaro, J.A., and Jarzynski, J. (1981). *Appl. Phys. Opt.* **20**, 2305.

Laude, J.P., Flamand, J., Gautherin, J.C., Lepère, D., Gacoin, P., Bos, F., and Lerner, J. (1983). *Eur. Conf. Opt. Commun.*, Geneva, p. 417.

Leslie, D.H., Trusty, G.L., Dandridge, A., and Giallorenzi, T.G. (1981). *Electron. Lett.* **17**, 581.

Lewis, F.A. (1967). *The Palladium–Hydrogen System*. Academic Press, New York.

Miller, S.E., Marcatili, E.A.J., and Li, T. (1973). *Proc. IEEE* **61**, 1973.

Narayanaswamy, R. (1987). *Proc. Symp. Chem. Sensors Electrochem. Soc.* **87-9**, 428.

Norris, J.O.W. (1991). In *Techniques and Mechanisms in Gas Sensing* (P.T. Moseley, J. Norris, and De Williams, Eds.), Chapter 11. Adam Hilger, Bristol.

Nylander, C. (1985). *J. Phys. E.* **18**, 736.

Othonos, A., Alavič, T., Melle, S., Karr, S., and Measures, R. (1993). *Opt. Eng.*, (in press).

Peterson, J.I., and Vurek, G.G. (1984). *Science* **244**, 123.

Peterson, J.I., Goldstein, S.R., and Fitzgerald, R.V. (1980). *Anal. Chem.* **52**, 864.

Seiler, M.R., and Leach, E.R. (1982). *J. Appl. Phys.* **53**, 5498.

Seitz, W.R. (1984). *Anal. Chem.* **56**, 16A.

Smela, E., and Santiago-Aviles, J.J. (1988). *Sens. Actuators* **13**, 117.

Szarka, F.H. (1988). *Fiber Integr. Opt.* **8**, 135.

Tien, P.K., and Capik, R.J. (1980). *Top. Meet. Integr. Guided-wave Opt.*, Nevada, p. TuB3-1

Wolfbeis, O.S., Weis, L.J., Leiner, M.J.P., and Ziegler, W.E. (1988). *Anal. Chem.* **60**, 2028.

Yariv, A., and Winsor, H.V. (1980). *Opt. Lett.* **5**, 87.

CHAPTER

6

PIEZOELECTRIC QUARTZ CRYSTAL MICROBALANCE SENSORS

6.1. INTRODUCTION AND HISTORICAL PERSPECTIVE

In 1880 Pierre and Jacques Curie observed for the first time that a pressure exerted on quartz resulted in an electric field between the two deformed surfaces (Maarsen et al., 1957). Conversely, the application of an electric field between the two surfaces caused a certain deformation. These two symmetric effects in some materials were named the piezoelectric effect by the two aforementioned distinguished French scientists. Etymologically the name of the phenomenon is derived from the Greek verb πιεζω (*piezo*), which means "exert pressure." As a result of the above discoveries, various crystals were used as piezoelectric materials during the first half of the twentieth century. These materials were employed for the manufacture of piezoelectric wave devices. In fact, piezoelectric crystals were first used as acoustic transducers as early as 1917 (Mason, 1950).

By 1964, almost five decades after the appearance of the first piezoelectric acoustic devices, new applications appeared in piezoelectric technology. These included the first gravimetric sensors to detect adsorption of gases (King, 1964, 1969, 1970, 1981). The foregoing author showed experimentally that a coated quartz piezoelectric crystal could be used as a sorption detector. Very quickly the new solid state gas sensor, a gas detection device, proved to be extremely useful and was established as the piezoelectric quartz crystal microbalance (PQCMB). The piezoelectric detector principle is well known and was described by Sauerbrey as early as 1959. Sauerbrey's (1959, 1964) suggestions were based on the application of the PQCMB device for measuring the thickness of thin films and showed that the shift in resonant frequency was within $\pm 2\%$. Sauerbrey's theoretical predictions were reconfirmed and extended experimentally by several workers such as Oberg and Lingenjo (1959), Behrndt and Love (1962), Warner and Stockbridge (1962), Stockbridge (1966), Eer Nisse (1967), Miller and Bolef (1986a,b), and finally by Lu and Lewis (1972). Remarkably, all of this work appeared between 1959 and 1972. However, up to that period use of the bulk piezoelectric devices was limited to only those researchers with in-house fabrication

179

facilities and research applications were produced by a small number of scientific groups around the world. By the mid-1970s the PQCMB was being developed with the use of highly sophisticated technology, optimized and tested in the silicon microelectronics area. At that stage the PQCMB device was introduced as a digital sensor in a complete digital system (Janghorbani and Freund, 1973). Later on, the array of piezoelectric systems was considerably improved and adapted to more advanced and sophisticated electronics (Carey and Kowalski, 1988). These sensor systems were shown to provide multivariable data on gas-phase mixtures of vapors (Carey et al., 1987). On the other hand, the parallel progress in the ultrahigh vacuum (UHV) sciences and the necessity of environmental monitoring led to the optimization of the device.

Today, highly sophisticated single, dual, and array piezoelectric-microsensor-controlled devices are available commercially. The manufacture of piezoelectric devices has made excellent progress, and thus new piezoelectric materials such as polymers (Lovinger, 1983) have appeared with a considerable reduction in cost. Polymeric materials are now used quite extensively in the manufacture of piezoelectric devices, and sensors in general. In Chapter 8, for instance, we describe the first active thin-film polymer sensor.

In 1984, Lu and Czanderna produced an excellent multiauthor book with details regarding practical applications of PQCMB (see Lu, 1984); the main areas of scientific research in bulk piezoelectric devices were synopsized as follows:

a. Thin-film deposition process control

b. Simultaneous mass and temperature measurements

c. Applications for the surface sciences, i.e., metal–gas systems

d. Plasma-assisted etching studies and applications

e. Space system contamination studies

f. Aerosol mass measurements

g. Analytical chemistry and gas detection

j. Detection of liquid phases

In this chapter our main purpose is to give a simple physical description of the bulk piezoelectric device as a gas sensor. We review the present level of understanding of these devices, their problems, and what their advantages may be. In Section 6.2 we present the current state of the theory, yet we do not try to expose the complete piezoelectric theory but rather we review and address only some key points in order to lead the reader to a better understanding of the operation of the device in gas detection analysis, rather than getting immersed in mathematical models. Section 6.3 is devoted to the experimental apparatus that shows the principle of the piezoelectric setup while providing some information concerning the availability of PQCMB devices. Section 6.4 is devoted to the piezoelectric detection of various gases. The subsequent sections present several experimental results and tables showing the types of coatings that have been used for various gas detection schemes as well as the influences of some interfering molecules on those PQCMB detections. In addition, several factors that influence the piezoelectric response such as type and thickness of coating, temperature, and flow rate will be discussed. The advantages and disadvantages are also critically reviewed. In Section 6.5 we present the technique and the use of an array of piezoelectric quartz crystal microbalance sensors (A–PQCMB) for multi-component gas analysis and we attempt to inform the reader about variations in these PQCMB systems. An extensive list of references is also included, as well as some informative tables concerning types of gases detected, chemical coatings, and gaseous interferences. In the last section of this chapter we present a table listing more general nongaseous applications of PQCMB devices.

6.2. THEORY OF THE PQCMB

6.2.1. Equivalent Electronic Circuit of the Quartz Crystal

It is well known that the piezoelectric crystal is simultaneously a condenser, a motor, and a generator that is dominated by a capacitive electronic behavior owing to its high resistivity (Mason, 1950). The piezoelectric quartz crystal resonator can be represented by the equivalent circuit of Fig. 6.1. The capacitance C represents the mechanical elasticity of the vibrating body; the inductance L is a measure of the vibrating mass of the crystal; the resistance R corresponds to the total loss of mechanical energy because of internal friction and dissipation of energy to the surrounding medium; C_0 is actually a lumped capacitance due to the electrodes on the quartz surfaces and the stray capacitance of the supporting mechanical structures. Resonance takes place in the unique situation where the complex impedance of the quartz crystal is purely resistive. In this case, there are two frequencies at which the resonant impedance is purely resistive and can be written as (Lu, 1984)

$$F_s = \frac{1}{2\pi} \frac{1}{(LC)^{1/2}}$$ (6.1)

$$F_p = \frac{1}{2\pi} \left[\frac{1}{LC} + \frac{1}{LC_0} + \left(\frac{R}{L} \right)^2 \right]^{1/2}$$ (6.2)

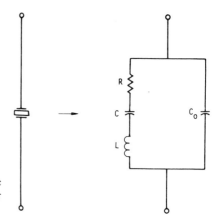

Figure 6.1. Schematic of the equivalent electronic circuit of a piezoelectric quartz crystal resonator (Lu, 1984).

where F_s and F_p are the series and parallel resonant frequencies, respectively. For typical piezoelectric quartz crystal resonators the following conditions hold (Lu, 1984):

$$\frac{1}{LC_0} \gg \left(\frac{R}{L}\right)^2 \quad \text{and} \quad \frac{C}{C_0} \ll 1. \tag{6.3}$$

The quartz crystal parallel resonant frequency, F_p, can be given approximately as

$$F_p = \frac{1}{2\pi \cdot (LC)^{1/2}}\left(1 + \frac{C}{2C_0}\right) = F_s\left(1 + \frac{C}{2C_0}\right) \tag{6.4}$$

The foregoing simple relation cannot be used for any quantitative analysis because the exact values of L and C in the equivalent circuit cannot be easily derived from the physical properties of the resonator. The equivalent electric circuit is only useful for designing crystal oscillator circuits and analyzing their operation.

6.2.2. Thermodynamic Considerations of the Quartz Crystal

According to the Onsager relations (Smith et al., 1967), a stress applied to the piezoelectric system generates charges and the introduction of a charge (electric field or voltage) will introduce a mechanical strain. This subsection will not describe the complete theory of the piezoelectric phenomenon since several other authors have already published a complete description of the piezoelectric model (e.g., see Zemel, 1985). Nevertheless, some important fundamental equations will be given. The Onsager relations are valid, of course, only in the case of the elastic regime where solid deformations are completely reversible. Thermodynamically, the deformation of a solid piezoelectric crystal by body and surface forces induces a change in the internal energy of the isolated system. As a result, the net work of the system is due to the local deformation of the solid by volume and surface forces as well as due to the variation of the charge density. Thus the total work can be written as

$$\delta W = -(\boldsymbol{\sigma}{:}\delta\boldsymbol{\Sigma} + \mathbf{E}{\cdot}\delta\mathbf{D}) \tag{6.5}$$

where $\mathbf{E}$ is the electric field and $\mathbf{D}$ the electric displacement vector; $\boldsymbol{\sigma}$ and $\boldsymbol{\Sigma}$ are the stress and strain tensors. The negative sign appears according to the standard convention to indicate work done by the thermodynamic system on the external agent. From the definition of the second law of thermodynamics (Landau and Lifschitz, 1959) and taking into account Eq. (6.5), one can write:

$$\delta U = T\,\delta S - \delta W = T\,\delta S + \boldsymbol{\sigma}{:}\delta\boldsymbol{\Sigma} + \mathbf{E}{\cdot}\delta\mathbf{D} \tag{6.6}$$

where T is the absolute temperature and S the entropy. The Gibbs free energy is given by

$$G = U - TS - \boldsymbol{\sigma}{:}\boldsymbol{\Sigma} - \mathbf{E}{\cdot}\mathbf{D} \tag{6.7}$$

Comparing Eq. (6.6) and the differential form of Eq. (6.7), one can write the Gibbs–Duhem equation:

$$\delta G = -S\,\delta T - \boldsymbol{\Sigma}{:}\delta\boldsymbol{\sigma} - \mathbf{D}{\cdot}\delta\mathbf{E} \tag{6.8}$$

By expanding the free energy up to linear terms, one obtains

$$\delta G(T, \mathbf{E}, \boldsymbol{\sigma}) \approx \left.\frac{\partial G}{\partial T}\right|_{\mathbf{E},\boldsymbol{\sigma}} \delta T + \left.\frac{\partial G}{\partial \mathbf{E}}\right|_{T,\boldsymbol{\sigma}} {\cdot}\delta\mathbf{E} + \left.\frac{\partial G}{\partial \boldsymbol{\sigma}}\right|_{\mathbf{E},T} {:}\delta\boldsymbol{\sigma} \tag{6.9}$$

and from Eqs. (6.8) and (6.9) the following definitions can be made:

$$-S = \left.\frac{\partial G}{\partial T}\right|_{\mathbf{E},\boldsymbol{\sigma}} \tag{6.10}$$

$$-\mathbf{D} = \left.\frac{\partial G}{\partial \mathbf{E}}\right|_{T,\boldsymbol{\sigma}} = \nabla_E G|_{T,\boldsymbol{\sigma}} \tag{6.11}$$

$$-\Sigma = \frac{\partial G}{\partial \sigma}\bigg|_{T,E} = \sum_{i,j} \frac{\partial G}{\partial \sigma_{i,j}}\bigg|_{T,E} \tag{6.12}$$

From these equations one can set up the usual Maxwell's relations and easily show that

$$\frac{\partial S}{\partial E} = \frac{\partial \mathbf{D}}{\partial T} \equiv \mathbf{p} \qquad \text{pyroelectric vector} \tag{6.13}$$

$$\frac{\partial S}{\partial \sigma} = \frac{\partial \Sigma}{\partial T} \equiv \alpha \qquad \text{thermal expansion tensor} \tag{6.14}$$

$$\frac{\partial \mathbf{D}}{\partial \sigma} = \frac{\partial \Sigma}{\partial E} \equiv \mathbf{d} \qquad \text{piezoelectric tensor} \tag{6.15}$$

In the linear regime of the forces and the responses of the sensor structure, one can write the incremental strain as a linear differential equation in δT, δE, and $\delta \sigma$ (DeGroot and Mazur, 1969):

$$\delta\Sigma \equiv \frac{\partial \Sigma}{\partial T}\bigg|_{E,\sigma} \delta T + \frac{\partial \Sigma}{\partial E}\bigg|_{T,\sigma} \cdot \delta E + \frac{\partial \Sigma}{\partial \sigma}\bigg|_{T,E} :\delta \sigma \tag{6.16}$$

The strain can be written as

$$d\Sigma = \psi : \delta\sigma + \mathbf{d} : \delta E + \alpha\, \delta T \tag{6.17}$$

where $\psi = \partial\Sigma/\partial\sigma|_{T,E}$. It can then be shown by expanding the displacement vector $\mathbf{D}(T, E, \sigma)$, up to the linear terms, that

$$\delta\mathbf{D} = \frac{\partial \mathbf{D}}{\partial T}\bigg|_{E,\sigma} \delta T + \frac{\partial \mathbf{D}}{\partial E}\bigg|_{T,\sigma} :\delta E + \frac{\partial \mathbf{D}}{\partial \sigma}\bigg|_{T,E} :\delta\sigma \tag{6.18}$$

By comparing Eqs. (6.11), (6.13), and (6.17) one can now write (Zemel, 1985):

$$\delta\mathbf{D} = \mathbf{p}\,\delta T + \epsilon :\delta E + \mathbf{d} :\delta\sigma \tag{6.19}$$

where ϵ is the dielectric tensor and is defined by

$$\frac{\partial \mathbf{D}}{\partial E}\bigg|_{T,\sigma} = \epsilon \tag{6.20}$$

Finally, the linear expansion of the entropy yields

$$\delta S = \frac{\partial S}{\partial T}\bigg|_{E,\sigma} \delta T + \frac{\partial S}{\partial E}\bigg|_{T,S} \cdot \delta E + \frac{\partial S}{\partial \sigma}\bigg|_{T,E} :\delta\sigma \tag{6.21}$$

which, upon use of thermodynamic definitions, results in

$$T\,\delta S = \rho_m C_p\,\delta T + T(\mathbf{p}\cdot\delta E + \mathbf{\alpha}:\delta\sigma) \tag{6.22}$$

where ρ_m is the mass density and C_p the specific heat of the crystal at constant pressure, derived from

$$\rho_m C_p = \frac{\partial S}{\partial T}\bigg|_{E,\sigma} \tag{6.23}$$

The foregoing fundamental equations contain the basic thermodynamics of operation of the piezo- and pyroelectric devices. In the former device $\delta\sigma$ plays the dominant role in the analysis, whereas δT is the analogous quantity for pyroelectric devices (as will be shown in Chapter 8).

6.2.3.   Wave Equation of the Piezoelectric Quartz Crystal

Establishing a wave equation for the piezoelectric structure is not a simple task owing to the tensorial relation between the piezoelectric and elastic phenomena. Nevertheless, as shown in Fig. 6.2(a), the surfaces perpendicular to the y and z axes are not subject to any stress. Because of the attachment of a conducting electrode to the two surfaces of the crystal perpendicular to the z axis, the field components in the y and z directions are zero. As a result, the stress tensor matrix can be written as follows:

$$\delta\sigma = \begin{bmatrix} \sigma_{xx} & 0 & 0 \\ 0 & 0 & 0 \\ 0 & 0 & 0 \end{bmatrix} \tag{6.24}$$

while the electric field vector can be written as

$$\delta E = \begin{bmatrix} 0 \\ 0 \\ \mathscr{E}_z \end{bmatrix} \tag{6.25}$$

Thus, it is easy to prove that the only force is along the x axis, and from

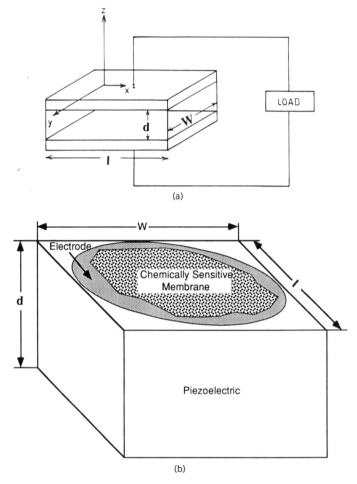

Figure 6.2. (a) Schematic layout of the piezoelectric structure. The applied electric voltage induces a piezoelectric oscillation around the resonance frequency (Zemel, 1985). (b) Illustration of a bulk piezoelectric crystal with the chemically sensitive membrane, which is the source of the response by absorbing or otherwise reacting with environmental vapors (Zemel, 1990).

Eq. (6.16) the strain component in the x direction is

$$\Sigma_{xx} = \psi_{11}\sigma_{xx} + d_{13}\mathscr{E}_z = \Sigma_{11} \tag{6.26}$$

where d_{13} is the nonzero term in the matrix of the piezoelectric tensor of the quartz crystal (Zemel, 1985). By taking into account the dielectric characteristics of the piezoelectric materials, one can define for the schematic of Fig. 6.2(a) the general wave equation (Zemel, 1985)

$$\mathbf{V} \cdot \mathbf{\sigma} = \frac{\partial \sigma_{xx}}{\partial x} = -\frac{1}{\psi_{11}}\left(\frac{\partial^2 u}{\partial x^2} - d_{13}\frac{\partial \mathscr{E}_z}{\partial x}\right) \tag{6.27}$$

Equation (6.27) is the product of Poisson's equation and Newton's second law. Usually $\mathscr{E}_z$ is a not a function of the lateral x position because the capacitive electrodes are equipotentials,

$$\frac{\partial \mathscr{E}_z}{\partial x} = 0 \tag{6.28}$$

Therefore, one can write the wave equation as

$$\rho_m \frac{\partial^2 u}{\partial t^2} = \frac{1}{\psi_{11}}\frac{\partial^2 u}{\partial x^2} \tag{6.29}$$

Equation (6.29) possesses the solution

$$u(x, t) = u(x)\exp(i\omega t) \tag{6.30}$$

where $\omega = 2\pi f$; here f is the frequency of the electric oscillation, and $i^2 = -1$. By applying the boundary conditions of the model, $\sigma_{xx} = 0$ at $x = 0$ and $x = l$, one can easily show that the amplitude of the longitudinal modes is given by the relation

$$u(x) = d_{13}\mathscr{E}_z\left[\frac{\sin[m(l-2x)/2]}{\cos(ml/2)}\right] \tag{6.31}$$

where l is the length of the piezoelectric crystal [see Fig. 6.2(a, b)] and $m = \omega/V$. From the simplified current equation in the case of piezoelectric materials for which the electrical conductivity is negligible in comparison with the displacement vector, and from Eq. (6.18), one can write

$$\mathbf{J} = -\frac{\partial \mathbf{D}}{\partial t} = -\left[\mathbf{\epsilon}\cdot\frac{\partial \mathbf{E}}{\partial t} + \mathbf{d}:\frac{\partial \mathbf{\sigma}}{\partial t}\right] = -i\omega(\mathbf{\epsilon}\cdot\mathbf{E} + \mathbf{d}:\mathbf{\sigma}) \tag{6.32}$$

the current along the z axis can be written as

$$J_z = -i\omega\left(\epsilon_z\epsilon_0\mathscr{E}_z + \frac{d_{13}}{\psi_{11}}\frac{\partial u}{\partial x}\right) \tag{6.33}$$

where ϵ_z is the diagonal component of the dielectric tensor in the z direction, and ϵ_0 is the permittivity of free space. By taking into account the wave equation solution given in Eqs. (6.31) and (6.33), the current density obtained by integration over the electrode surface is

$$I = \int_0^l J_z(x)W\,dx = -i\omega W l\left[\epsilon_z\epsilon_0 + \frac{d_{13}^2 \tan(ml/2)}{ml/2}\right]\mathscr{E}_z \qquad (6.34)$$

where W is the width of the piezoelectric [Fig. 6.2(a,b)]. The impedance ζ of the circuit can be written as

$$\frac{1}{\zeta} = \frac{i\omega A}{d}\left[\epsilon_z\epsilon_0 + \frac{d_{13}^2}{\psi_{11}} \frac{\tan(ml/2)}{ml/2}\right] \qquad (6.35)$$

where $A = W \times l$. When $\tan(ml/2) = \infty$ the impedance vanishes, a condition corresponding to a resonance of the circuit of Fig. 6.1. Thus, the resonance frequency F arises from the condition (Zemel, 1985)

$$\frac{ml}{2} = -(2n+1)\frac{\pi}{2} \qquad (6.36)$$

The first harmonic ($n = 0$) is given by

$$\frac{ml}{2} = -\frac{\pi}{2} \qquad (6.37)$$

From the velocity of sound, V, in dielectrics one can express the resonant frequency, F, with the relation

$$F = \frac{V}{2l} = \frac{1}{2l}(\rho\psi_{11}) \qquad (6.38)$$

In the case of PQCMB, F is on the order of 5–15 MHz. The resonant frequency can also be expressed in terms of the total mass, M, of the piezoelectric crystal and the total thickness, d, of the quartz in Fig. 6.2. If one replaces the mass density ρ by M/V, the resonance frequency is given by the fundamental relation (Guilbault, 1984)

$$F = \frac{1}{2}\left(\frac{1}{ldW\Sigma_{11}}\right)^{1/2} \qquad (6.39)$$

where $\Sigma_{11} = \psi_{11}\sigma_{11}$.

6.2.4. Sauerbrey's Theoretical Approach

The oscillation frequency of a quartz crystal depends on the total mass of the crystal and that of any coating layers (or electrodes) on the crystal surfaces. When adsorbed gas molecules are absorbed in the thin coating layer, the resonant frequency decreases in proportion to the mole number of dissolved molecules. Thus, the concentration of a pollutant gas is measured by detecting a change in the crystal vibration frequency. The accuracy of gas detection and monitoring depends upon the sensitivity of the frequency change to shifts caused by factors other than mass, such as instrumental effects, temperature fluctuations, stress, and the accuracy of the mathematical model used to convert the resonant frequency change, ΔF, to mass addition.

It is important to note that in many dynamic mass measurements, such as in kinettic studies, the rate of mass change is also of high interest. Thus the evolution of ΔF with time can also be monitored by combining the mass change and the kinetics that take place. Until 1950 the frequency shift, ΔF, was a phenomenon that was described only qualitatively. The necessity for monitoring small mass changes later led researchers to more careful investigations. By 1960, in fact, it was already known that the resonant frequency of a quartz crystal was dependent on the geometrical dimensions of the quartz plate and the thickness of its electrodes. Therefore, manufacturers used to prepare quartz crystals with resonant frequencies higher than the desired value and then reduced F to a more precise value by controlling the thickness of the deposited quartz electrodes (Lu, 1984).

The theory outlined below was originally developed by Sauerbrey (1959, 1964). Figure 6.3 shows a simplified model of a PQCMB. A quartz crystal plate oscillating in the fundamental thickness–shear mode must satisfy the equation

$$z_q = \frac{\lambda_q}{2} \tag{6.40}$$

where z_q is the quartz plate thickness, and λ_q is the wavelength of the shear-mode elastic wave in the thickness direction. The shear wave velocity V_q is given by

$$V_q = \lambda_q F \tag{6.41}$$

By combining Eqs. (6.40) and (6.41) one can write

$$F z_q = \frac{V_q}{2} \tag{6.42}$$

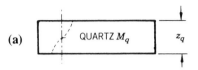

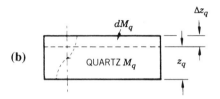

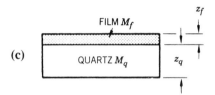

Figure 6.3. Schematic of a simplified model of a PQCMB: (a) At resonance the wavelength is equal to half the thickness of the quartz plate. (b) The increase of the thickness of the quartz results in a decrease of the resonant frequency. (c) The mass of a deposited film is treated as an equivalent amount of mass of the quartz crystal (Lu, 1984).

The resonant frequency shift caused by an infinitesimal change in the crystal thickness Δz_q is found to be

$$\frac{\Delta F}{F} = -\frac{\Delta z_q}{z_q} \qquad (6.43)$$

where the negative sign indicates that an increase in the thickness of the quartz crystal plate causes a decrease in its resonant frequency F. Owing to this linear relation of the quartz crystal thickness and mass, one can express Eq. (6.43) in terms of quartz crystal mass M_q and its mass change ΔM_q:

$$\frac{\Delta F}{F} = -\frac{\Delta M_q}{M_q} \qquad (6.44)$$

Sauerbrey's fundamental assumption was that for small mass changes the addition of foreign mass can be treated as an equivalent mass change of the quartz crystal itself. Thus Eq. (6.44) can be written as

$$\frac{\Delta F}{F} = -\frac{\Delta M}{M_q} \qquad (6.45)$$

where now ΔM is an infinitesimal amount of foreign mass uniformly distributed over the crystal electrode surface owing to the absorption of gas molecules. Since the mass of quartz crystal is $M_q = A \times \rho \times z$, Eq. (6.45) can also be written as

$$\frac{\Delta F}{F} = -\frac{\Delta M}{A\rho z} \qquad (6.46)$$

where A is the total surface area (cm^2), and ρ is the density of quartz ($= 2.6\, g\, cm^{-3}$) (Kindlund and Lundström, 1982/83). For AT-cut[1] quartz one finds (Sauerbrey, 1959)

$$\Delta F(\text{Hz}) = -2.3 \times 10^6 F^2 \frac{\Delta M}{A} \qquad (6.47)$$

In this case F is the resonant frequency (MHz) of the crystal in the absence of gas absorption; ΔM is the mass of the absorbed gas molecules (g). According to Kindlund and Lundström (1982/83), the total mass before any absorption of gas is

$$M = M_{\text{PC}} + M_{\text{Pd}} \qquad (6.48)$$

where M_{PC} is the mass of the piezoelectric crystal, and M_{Pd} is the mass of the chemically sensitive layer (Pd in this case). The mass of the electrode-coated quartz crystal is approximately between 50 and 150 mg, and thus the resonant frequencies, F, are between 5 and 12 MHz. Under these conditions, Eq. (6.47) shows that the minimum detectable mass of absorbed gas by the Pd layer is $\Delta M \approx 10^{-8}$ g. The theoretical detection limit for the piezoelectric quartz crystal detector appears not to have been calculated rigorously from first principles. Sauerbrey (1964) estimated this limit to be approximately 10^{-12} g. King's estimation (1964) was approximately 10^{-9} g. It is obvious from Eq. (6.47) that mass sensitivity is better for detectors with high F. Equation (6.47) can also be expressed as a function of concentration (Shackleford and Guilbault, 1974; Bastiaans, 1988):

$$\Delta F = -k_c \Delta C_0 \qquad (6.49)$$

where k_c is a constant related to the resonant frequency of the coated quartz plate, and it includes a conversion factor between the mass of the absorbed

[1] An AT quartz plate is a plate cut from the natural crystal perpendicular to the x axis and at $35°15'$ from the z axis.

gas (g) and the concentration ΔC_0 (ppm or %) of the gas in the gas phase. The linear relation of Eq. (6.49) with respect to the gas concentration is an important advantage of the piezoelectric technique.

6.3. PIEZOELECTRIC EXPERIMENTAL INSTRUMENTATION

There are several types of piezoelectric experimental setup described in the literature. Generally, each research group has sought to design a system adapted to its special experimental conditions. For example, different priorities of system construction are given in the case of a flow dynamic study as opposed to that requiring static measurements. The temperature conditions and the working ambient pressure are also very important factors one has to take into account when putting together an experimental setup. In fact, PQCMB detectors working under UHV conditions require different (supplemental) instrumentation from a PQCMB under STP conditions.

Nevertheless, all the various piezoelectric systems that have been constructed during the last 30 years contain four important subsystems: gas control, temperature control, test cell, and shift frequency analysis instrumentation (Fig. 6.4).

The *temperature control system* ensures that the temperature of the incoming gas flow (in the case of dynamic measurements) will be the same as that of

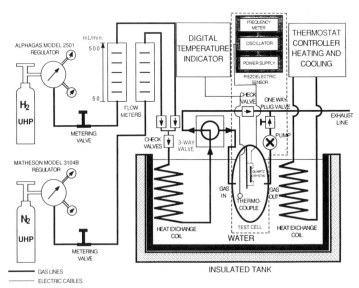

Figure 6.4. Schematic diagram of a generic PQCMB experimental system (Christofides and Mandelis, 1989).

the sensor itself. A heat exchange coil is installed on the gas mixture line, upstream from the test cell. The test cell and heat exchange coil are suspended in a water bath to equalize their temperatures. The effect of temperature on PQCMB sensors is very important: (a) the quartz crystal shows a frequency shift with temperature change (Shackleford and Guilbault, 1974), and (b) it is well known that the temperature influences the gas–surface (i.e., coating) interactions quite markedly (Lewis, 1967; Lundström et al., 1989).

The *gas control system* mixes the gases in a homogeneous flow. The flow rate of each gas can be adjusted and stabilized before the mixture is directed into the test cell. The system includes tanks of gases, pressure regulators, flow meters, valves, a pump, and manifold lines that interconnect all of these components. Figure 6.4 shows the complete experimental setup used in a typical PQCMB study. The gas-handling system introduces a mixture of gases into the test cell and allows the mixture to leave through the exhaust line after pumping.

A working experimental setup, as in Fig. 6.4, may be designed to ensure that the temperature of the incoming gas flow will be the same as that of the sensor itself. A metallic disc may be installed in front of the gas inlet, in order to avoid acoustic noise by dispersing the incoming gas to different directions in the test cell, thus obtaining a homogeneous stream. The heat exchange coil further helps in homogenizing the gas mixture. Several thermocouples may be installed inside the test cell to monitor the background temperature, especially the PQCMB performance in the presence of temperature gradients as the thermal bath temperature is scanned during diagnostic measurements of the device. Finally, the use of thermal insulation around the tank to obtain stable signals over long periods of sampling time is often recommended.

The *test cell* is the heart of the experimental system. It consists of a pressure vessel containing the piezoelectric quartz crystal detectors. An experimental setup of the control piezoelectric quartz crystal detector used by Abe and Hosoya (1984) is shown schematically in Fig. 6.5.

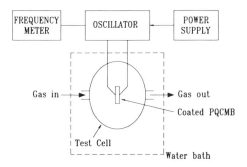

Figure 6.5. Electronic setup of the PQCMB detector.

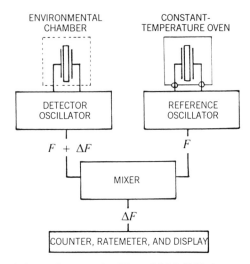

Figure 6.6. Dual electronic setup of a differential PQCMB detector (Zemel, 1985).

The piezoelectric instrumentation consists of an oscillator powered by a regulated power supply, ensuring that the applied voltage is kept constant (10–20 V dc). The frequency output from the oscillator is measured with a frequency counter having a high resolution (1–0.1 Hz). The response of the PQCMB sensor can be observed as a change in the resonant frequency of the crystal in agreement with the simple theoretical considerations presented above (Sauerbrey, 1959, 1964).

Figure 6.6 shows a dual electronic setup of a differential PQCMB detection system. Such systems present higher sensitivity than the single-mode detection of Fig. 6.5. In fact, one of the piezoelectric crystals in the case of the dual system is coated with the appropriate chemically sensitive layer and the second quartz possesses neutral electrodes. It serves to correct frequency variations due to environmental changes (i.e., temperature, pressure, etc.). In this case F_A and F_R are the resonant frequencies of the active and reference quartz crystals, respectively, and in the presence of gas the differential frequency change can be written as

$$\delta F = F_A - \Delta F_A - F_R \tag{6.50}$$

6.4. PQCMB FOR GAS DETECTION: EXPERIMENTAL RESULTS

During the last four decades, coated piezoelectric crystals have demonstrated good efficiency as detectors of various pollutants. The piezoelectric micro-

Table 6.1. Type of Cases Detected by a PQCMB

References	Detected Gas
Frazier and Glosser (1979)	Hydrogen
Mecea and Ghete (1984)	Hydrogen
Abe and Hosoya (1984)	Hydrogen
Christofides and Mandelis (1989)	Hydrogen
Lopez-Roma and Ghuilbault (1972)	Sulfur dioxide
Frechette and Fasching (1973)	Sulfur dioxide
Karmarkar and Guilbault (1974)	Sulfur dioxide
Guilbault and Lopez-Roman (1975)	Sulfur dioxide
Karasek and Tiernay (1974)	Sulfur dioxide
Cheney and Homolya (1975)	Sulfur dioxide
Karmarkar et al. (1976)	Sulfur dioxide
Cheney and Homolya (1976)	Sulfur dioxide
Cheney et al. (1972)	Sulfur dioxide
Karmarkar and Guilbault (1975)	Ammonia
Hlavay and Guilbault (1978a)	Ammonia
Webber and Guilbault (1976)	Ammonia
Karmarkar and Guilbault (1975)	Nitrogen dioxide
NASA Tech. Brief (1968)	Hydrogen chloride
Hlavay and Guilbault (1978b)	Hydrogen chloride
Webber et al. (1978a, b)	Hydrogen sulfide
Shackleford and Guilbault (1974)	Hydrocarbons
Kindlund and Lundström, 1982/83	Hydrocarbons
Karmarkar et al. (1975)	Hydrocarbons
Scheide and Guilbault (1972)	Organophosphorus
Guilbault and Tomita (1980)	Organophosphorus
Guilbault et al. (1981)	Organophosphorus
Guilbault et al. (1981/82)	Organophosphorus
Kristoff and Guilbault (1983)	Organophosphorus
Janghorbani and Freund (1973)	Mercury
Scheide and Taylor (1974)	Mercury
Tomita et al. (1979)	Mercury
Lee et al. (1982)	Humidity
Randin and Zulling (1987)	Humidity
Ho et al. (1980)	Toluene
Guilbault et al. (1980)	Toluene
Tomita et al. (1979)	Mononitrotoluene
Guilbault and Ho (1982)	Carbon monoxide
Cooper et al. (1979)	Anesthetic vapors
Cooper et al. (1981)	Anesthetic vapors

balance device continues to be of interest to a goodly number of scientific groups around the world. Table 6.1 gives a global review of PQCMB devices and lists references including the type of gases which have been detected all these years.

It is interesting to note that Deakin and Byrd (1989) used a coated quartz crystal for the detection of electro-inactive cations in aqueous solutions. The effect of air and gas pressure on the vibration of miniature quartz tuning forks was investigated by Christen in 1983. The PQCMB has been used for several other applications; however, this chapter will only consider those applications concerning gas detection.

In this section, we shall present some of the most important developments concerning the detection of hydrogen in a flow-through system by a Pd-coated PQCMB. The influence of system parameters that affect the physicochemical mechanism of the sensor response is described. Some drawbacks of the piezoelectric detector are also discussed. Piezoelectric detection of hydrogen is considered first for the following reasons: (a) the growing importance of hydrogen gas; (b) the simplicity of the hydrogen molecule; (c) the Pd–H_2 system bridges academic and applied interests; and (d) hydrogen has been detected by most of the sensors presented in this book and thus can be a good common reference for a general critical comparison of gas sensors.

6.4.1. Hydrogen Bulk Piezoelectric Detection

A literature survey clearly shows that the science and technology of the detection of hydrogen by the piezoelectric quartz crystal coated with Pd electrodes has not progressed quite as rapidly as other detection schemes mentioned earlier. It is likely that the complex mechanisms of the hydrogen–palladium system and the strong interference of oxygen–Pd reactions (Lewis, 1967; Lasher, 1937; Everett and Nordon, 1960; Simons and Flanagan, 1965a,b; Ponec et al., 1969; Llopis, 1968; Vannice et al., 1970; Sermon, 1972) have made the quantitative aspects of such investigations very problematic. Frazier and Glosser (1979) have used a quartz crystal in order to determine the quantity of hydrogen absorbed by an evaporated palladium film. Bucur et al. (1976a) have used a PQCMB at 80.3 °C to study the kinetics of hydrogen sorption by thin Pd layers. The mechanism of hydrogen sorption by thin Pd layers has also been studied at a temperature of 61.1 °C (Bucur et al., 1976b). Under UHV conditions, Bucur (1981) has further used a PQCMB in order to study the effect of CS_2 molecules on the desorption kinetics of hydrogen from a thin Pd layer at 59.6 °C. Measurements for H_2–O_2 reactions using the piezoelectric quartz crystal under UHV conditions have also been performed (Kasemo and Tornqvist, 1980). Mecea and Bucur (1980) have used the piezoelectric device for sorption studies under dynamic conditions. Abe and Hosoya (1984) have used a piezoelectric quartz crystal for the detection of flowing hydrogen both in ambient air and in nitrogen (0.5% H_2 in N_2 and in air). Their limited study, however, did not lead to definite quantitative conclusions as to the hydrogen detection capabilities and operating mechanisms of the Pd–PQCMB sensor.

6.4.1.1. Influence of the Pd Thickness

It is important to point out that the thickness of the Pd coating/electrode plays an important role in the operation of the detector. Frazier and Glosser (1980) have shown the influence of the Pd thickness on the piezoelectric measurements in the range of 60–1200 Å at 27 °C. These authors also showed that the phase transition of the Pd depends strongly on its thickness. Figure 6.7 shows the reduced pressure–concentration (P–C) absorption isotherms for Pd hydrides thus obtained.

6.4.1.2. Sensitivity and Limitations

We shall now describe some experiments showing difficulties in the detection of hydrogen by a PQCMB at room temperature under flow-through conditions (Christofides and Mandelis, 1989, 1990; Abe and Hosoya, 1984). It has been realized (Abe and Hosoya, 1984) that the pretreatment of the detector prior to any experiment plays an important role in the interpretation of the data. Initially, a Pd–PQCMB sensor was left for several days under ambient

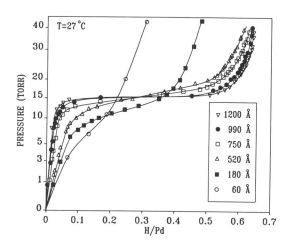

Figure 6.7. Pressure–concentration isotherms for Pd hydride for various thicknesses.

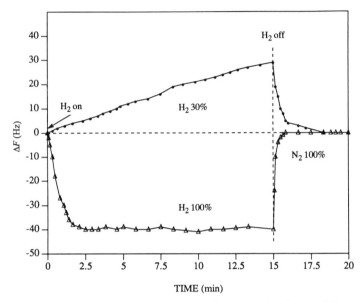

Figure 6.8. Piezoelectric quartz crystal responses as a function of time for two different concentrations of hydrogen in nitrogen: 30% and 100%. The test cell was exposed for several days to the laboratory ambient air before the first experiment ($F = 6$ MHz; Pd thickness: 800 Å; flow rate: 500 mL/min; $T = 20\,^\circ$C) (Christofides and Mandelis, 1989).

conditions in order to observe the influence of atmospheric air on the catalyst (Christofides and Mandelis, 1989). Afterward, hydrogen ($[H_2] = 30\%$ in N_2) was introduced into the test cell. Figure 6.8 shows the variation of the resonant frequency (an *increase* in ΔF) as a function of time. A signal decay toward the baseline was observed when the inlet valve was closed and the test cell evacuated, as expected. The phenomenon of a positive ΔF shift upon H_2 exposure is surprising in light of the simple accepted theory presented in Chapter 2 (Section 2.4.2); however, this shift has been observed by both Christofides and Mandelis (1989) and Abe and Hosoya (1984), who used 0.5% flowing gas mixtures of hydrogen in air and nitrogen. Figure 6.9 shows some of the experimental results indicating the increase of ΔF after adsorption–absorption of hydrogen (Abe and Hosoya, 1984). It has been hypothesized that the ambient oxygen plays an important role in the Pd–PQCMB response: preadsorbed O_2 on the Pd surface reacts with the introduced H_2 gas and forms H_2O which leaves the Pd surface via evaporation; therefore, the weight of the quartz crystal decreases as a result of the H_2–O_2 reaction, which is consistent with a positive ΔF. Vannice et al. (1970) have studied the absorption of H_2 and O_2 on platinum black. The great similarity between

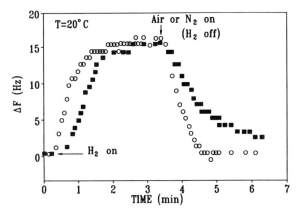

Figure 6.9. Piezoelectric quartz crystal responses as a function of time ($F = 6$ MHz; Pd thickness: 1500 Å; flow rate: 266 mL/min): (■) 5120 ppm hydrogen in nitrogen; (○) 5120 ppm hydrogen in air.

Pt and Pd as regards the absorption of H_2 and O_2 in these metals allows a reasonable comparison to be made between the results by Vannice et al. (1970) and the hypothesis put forth by Abe and Hosoya (1984), as well as the observations by Christofides and Mandelis (1989). It appears that the water formed during the reaction leaves the surface and, in doing so, is not replaced by additional hydrogen. All the empirical chemical reactions of the possible mechanism on the Pd surface with H_2 and O_2 dissociations and recombinations have been described in Chapter 2 (Section 2.4.2). However, according to Lundström et al. (1989), the details of the water production reactions on the palladium layer are still not known. This phenomenon turned out to be a significant disadvantage of the piezoelectric quartz crystal detector. The results obtained with the 30% H_2 in N_2 (Fig. 6.8) indicate that gas impurities in the flowing gas mixture, notably oxygen, may play the dominant role in the response of the sensor.

The relationship between gas concentration and Pd–PQCMB detector response, ΔF, is an important parameter for sensor characterization. Figure 6.10(a) shows the variation of saturation resonant frequency, ΔF_s, as a function of gas phase hydrogen concentration (Christofides and Mandelis, 1989). Unlike the prediction of Eq. (6.49), there are three apparent response regimes; a nonlinear one at concentrations between 20% and 70%; two linear regimes at low ($< 20\%$) and high ($> 70\%$) concentrations; and a region where all experimental curves converge. This convergence of all the experimental data can be explained by assuming that at high pressures the adsorbed and absorbed hydrogen dominates all other interfering phenomena. The presence

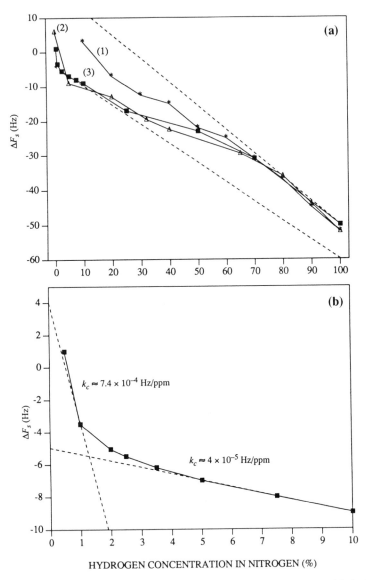

Figure 6.10. (a) Variation of the saturation resonance frequency shift, ΔF_s, with hydrogen concentration (or partial pressure). The three curves were obtained from the original exposure where the presence of ambient air before the onset of the experiment was significant: ΔF_s of curve (1) (unpurged) is consequently smaller than that of curves (2) and (3), where the cell was purged with N_2 before the experiment. (b) Variation of ΔF_s as a function of hydrogen partial pressure and concentration of hydrogen in nitrogen in the range of 0.1–10% hydrogen in N_2 (Christofides and Mandelis, 1989).

of trace oxygen is thought to be partly responsible for the deviation of the Pd–PQCMB sensor response at low concentrations. Figure 6.10(a) indicates that the purging history of the test cell plays an important role in determining the $\Delta F_s([H_2])$ curve at lower ($\leqslant 70\%$) concentrations. An important factor is the initial irreversible sensitization of the Pd–PQCMB surface with increasing degree of exposure to hydrogen flow. Figure 6.10(b) shows the variation of the saturation, ΔF_s, as a function of gas phase hydrogen concentration (Christofides and Mandelis, 1989) in the range of 0.1–10% H_2 in N_2. Unlike the response predicted by Eq. (6.49), there are two apparent linear response regions: one with the slope $k_c \approx 7.4 \times 10^{-4}$ Hz/ppm at low concentrations ($< 2\%$), and another at higher concentrations (3–10%) with the slope $k_c \approx 4 \times 10^{-5}$ Hz/ppm. The presence of trace oxygen even after the evacuation of atmospheric air with the help of N_2 seems to be partly responsible for the deviation of the Pd–PQCMB sensor response and the positive ΔF_s at low concentrations (Christen, 1983; Abe and Hosoya, 1984). In fact, at very low hydrogen concentrations many impurities such as oxygen may still exist on the Pd surface even after several activation cycles. On the other hand, for higher H_2 concentrations, a clean surface area will be generated owing to the reduction of the oxygen layer. The anomalous behavior at low concentrations—between 20% and 70% [see Fig. 6.10(a), curves (2) and (3)]—may be due to the $\alpha \rightarrow \beta$ phase transition that takes place around 20 kPa. According to Lundström et al. (1989), it is probable that a phase transition from α to β may also take place at low temperatures and high concentrations (1 to 100%) (Steele et al., 1986). This is in disagreement with other evidence (Lewis, 1967) showing the possibility that at room temperature the phase transition takes place around 2% (2 kPa). It is also well known that the phase transition depends on the palladium thickness (Frazier and Glosser, 1980), a complicating factor. Another problem for a quantitative analysis may be the presence of water on the Pd layer at room temperature. As was pointed out by Lundström et al. (1989), this effect could cause the Pd layer to be saturated at low pressures and has been observed in studies under UHV conditions.

6.4.1.3. Influence of Temperature and Flow Rate

The temperature variation during the gas detection by the Pd–PQCMB plays an important role because of the influence of temperature on the quartz crystal operation. According to Sauerbrey (1959), if a piezoelectric AT-cut or BT-cut[2] crystal undergoes a temperature change, a significant variation of the resonant frequency, F, may be expected. Abe and Hosoya (1984)

---

[2]A BT quartz crystal plate is a plate cut from the natural crystal perpendicular to the x axis and at $-49°$ from the z axis.

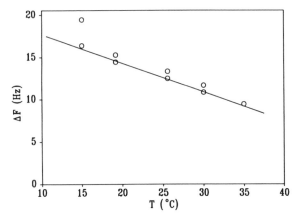

Figure 6.11. Effect of temperature on the sensitivity of the Pd–PQCMB device under exposure to 5120 ppm H_2 in N_2. Overlapping data points correspond to experimental reproducibility scatter.

reported a variation of 0.75 Hz/°C in the range of 15–30 °C: Figure 6.11 presents the effect of temperature on the resonant frequency of a Pd–PQCMB device.

Christofides and Mandelis (1989) have also studied the effect of flow rate, W_r, on sensitivity, and response time was measured using 5% of hydrogen in nitrogen. The flow rate was varied from 60 to 500 mL/min. It was observed that W_r does not influence ΔF_s very much, and saturation occurs essentially at the same level. On the other hand, the response time, R_s, of the sensor increases monotonically (7–14 min) with a decreasing flow rate between 500 and 60 mL/min, as shown in Fig. 6.12. We note the R_s is strongly dependent on W_r, especially at the low rates. At higher flow rates (> 400 mL/min), the response time becomes essentially independent of flow rate value. This observation was made earlier by Cooper et al. (1981). According to these authors the response time, R_s, is proportional to V_c/W_r, where V_c is the volume of the test cell. The experimental results on Fig. 6.12 do not follow a $1/W_r$ law. Cooper and co-workers (1979, 1981) attributed a similar deviation to the bad mixing of the detectable and carrier gases. The foregoing results suggest that minimizing the test cell volume offers optimal responses of the Pd–PQCMB sensor.

Before concluding this subsection we should note that another variation of the PQCMB has been developed and used by Mecea and Ghete (1984) in order to detect hydrogen. These authors have shown that a compact hydrogen detector utilizing an ultrasonic resonance within a cavity results in faster response and recovery times. In fact, the detection of 0.68% and

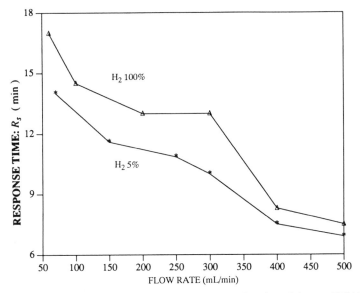

Figure 6.12. Response time of the Pd–PQCMB device as a function of time at 100% H_2 and 5% H_2 in N_2 for various flow rates at room temperature. The hydrogen was turned on at $t = 0$ min (Christofides and Mandelis, 1989).

3.9% of hydrogen in air requires only 24 and 4.3 s, respectively. For the same concentrations, only 4.2 and 11.8 s, respectively, are needed for complete recovery.

6.4.2. Halothane and Other Hydrocarbon Bulk Piezoelectric Detection

Kindlund and Lundström (1982/83) have extensively studied the detection of halothane by using a PQCMB sensor. These authors investigated the suitability of silicon oils as coatings for the detection of hydrocarbons. Several parameters such as stability, response, and recovery times, interferences, as well as the heat of adsorption and temperature dependence of the silicon-oil-based sorption detectors, were checked closely for several hydrocarbons. The nonselectivity of the silicon-oil coatings with respect to the various types of hydrocarbons has also been shown. Table 6.2 shows several types of coatings that have been used by several experimental groups for the detection of various gases. It has also been proven that the oil absorbs water molecules; however, this absorption can be distinguished from hydrocarbon absorption owing to the different heats of adsorption.

Kindlund and Lundström (1982/83) used a quartz crystal of 8.9 mm diameter with a resonant frequency of 12 MHz. The detector was coated

Table 6.2. Types of Chemical Coatings on PQCMB for Various Vapors and Gases: Detection Abilities

Detected Gas	Typical Coating	Sensitivity Level
H_2	Palladium	ppm
SO_2	Quadrol	ppb
	Triethanolamine	ppb
	Armeen 2S	ppm
H_2O	LiCl	ppm
	SiO_x	
NH_3	Ucon-75-H-90,000	ppb
	Ucon-LB-300X	
	L-Glutamic acid	
	Ascorbic acid	
	$AgNO_3$	
H_2S	Acetone	ppm
	Metallic copper	
HCl	Tertiary amines	ppm to ppb
	Triphenylamine	
Pesticide	Triton X-100	ppm
	NaOH	
Hydrocarbons	Silicon oil	ppm
Aromatic hydrocarbons	Nujol + Ir complex	ppm
Diisopropylmethyl phosphonate	$FeCL_3$ Paraoxon	ppm
Toluene	Carbowax	ppm
Mercury	Gold	ppm
Mononitrotoluene	Ucon-XMP-1018 Carbowax 1000	ppm
CO	HgO	ppb

with silicon oil and nitrogen gas used as a carrier. Various flow rates were utilized and temperature studies were performed. The main results of the PQCMB detection of hydrocarbons can be summarized as follows:

a. Influence of Flow Rates. At high flow rates the rate of frequency change of gas absorptions approaches a flow-independent value, as discussed earlier.

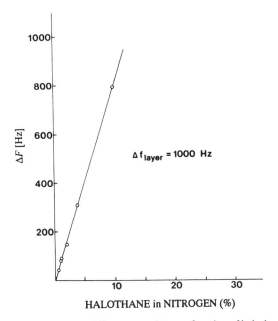

Figure 6.13. Piezoelectric frequency shift response, ΔF, as a function of halothane concentration in nitrogen at room temperature. The quartz crystal was coated with a silicon oil (Kindlund and Lundström, 1982/83).

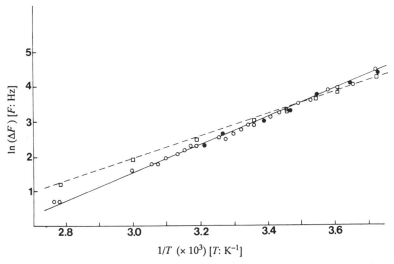

Figure 6.14. Temperature dependence of the Δf_{gas}, shown as $\ln|\Delta F|$ vs. $1/T$, for (○) 0.25 vol% halothane; (●) 0.5 vol% halothane with Δf_{gas} divided by 2; (□) $\ln(\Delta f_{gas})$ compensated for temperature dependence of the prefactor of the equilibrium constant. Here $\Delta f_{layer} = 1000\,Hz$; carrier gas = nitrogen (Kindlund and Lundström, 1982/83).

b. Detection Coatings. Fatty acids, paraffins, lecithins, and silicon oils were tried as sensitive layers. It was found that the silicon oils were superior regarding stability, sensitivity, and time response.

c. Linearity and Times of Response. The linear Eq. (6.49) was found to be satisfied up to 10% in the case of PQCMB detection of halothane at ambient room temperature. Figure 6.13 shows the linear frequency change vs. halothane concentration. The response and recovery times were found to be a fraction of a second.

d. Temperature Dependence and Energy of Adsorption. Figure 6.14 shows the variation of the $\ln|\Delta F|$ as a function of the inverse of temperature $1/T$ for a halothane concentration of 0.5% in nitrogen. This study was performed between -30 and $50\,°C$. The two curves shown in Fig. 6.14 were obtained from different experimental cycles. From the slope of the solid curve Kindlund and Lundström (1982/83) concluded that the absorption of halothane is an exothermic reaction with a heat of adsorption, E_{ads}, close to 0.35 eV/molecule. The aforementioned authors used a standard thermodynamic relation describing gas–surface interactions (Daniels and Alberty, 1966):

$$\Delta F \propto \frac{1}{m_g^{3/2}} \frac{1}{T^{5/2}} \exp\left(\frac{E_{ads}}{kT}\right) \tag{6.51}$$

where m_g is the mass of the gas molecule; k, Boltzmann's constant; and T, the absolute temperature. From the slope of the dashed line of Fig. 6.14, the foregoing authors found $E_{ads} = 0.28$ eV/molecule.

6.4.3.　Sulfur Dioxide Bulk Piezoelectric Detection

Several papers have been published describing the use of a PQCMB as a highly sensitive detector for sulfur dioxide (see Table 6.1). Guilbault (1984) in a review chapter has presented an extensive table of coating materials used to detect SO_2. Karmarkar and Guilbault (1974) employed triethanolamine and quadrol as coatings, enabling them to detect SO_2 in nitrogen even at ppb levels. Figure 6.15 shows the piezoelectric response to SO_2 with three different coatings: triethanolamine, quadrol, and Armeen 2S. It has been found that triethanolamine and quadrol are more sensitive to SO_2 than is Armeen 2S.

Guilbault and Lopez-Roman (1975) showed the influence of temperature on ΔF. Moreover, Cheney and Homolya (1975, 1976) investigated the adsorption, absorption, and desorption of SO_2 on the coating materials at various temperatures.

It has also been found that the deposition technique of the coating on

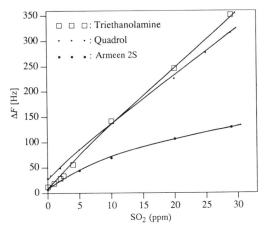

Figure 6.15. Piezoelectric response as a function of $[SO_2]$ concentration for three different chemical reactions (Karmarkar and Guilbault, 1974).

the crystal plays an important role on the response of piezoelectric devices. Earp (1966) has shown that the sensitivity of the sorption is inversely proportional to the distance from the center of the quartz crystal.

6.4.4. Detection of Other Gases

Table 6.2 presents the appropriate chemical coatings for the detection of several gases, as well as the sensitivity levels. In this subsection we shall highlight a few important aspects that have been discovered during the piezoelectric detection process of various gases such as ammonia, hydrogen sulfide, hydrogen chloride, organophosphorus and pesticides, aromatic hydrocarbons, mercury in air, mononitrotoluene, carbon monoxide, and others. The following summary of key observations can be made:

a. According to Hlavay and Guildbault (1978b), 7 min are needed for detection of ammonia.

b. Some 30 s were enough to detect aromatic hydrocarbons with Carbowax 550.

c. A small portable PQCMB device 20 cm × 14.7 cm × 9 cm has been used to detect 1–200 ppm of toluene.

d. For the detection of carbon monoxide an optimum temperature of 110 °C exists that helps the reaction on the coating of the quartz

(Guilbault and Ho, 1982):

$$HgO + CO \rightarrow Hg + CO_2 \qquad (6.52)$$

e. In the case of detection of mercury in air, reversibility can only be achieved by placing the detector in an oven at 50 °C. Scheide and Taylor (1974, 1975) have designed excellent mercury sensors as air pollution detectors for industrial hygiene purposes.

f. Figure 6.16 presents the linear response of the PQCMB as a function of several concentrations of organic vapors/gases, in agreement with Eq. (6.49).

g. In general, a lack of absolute selectivity during the PQCMB detection of gases may cause interference. Guilbault (1984) has presented several tables concerning these interferences. Some of these results are also included here for the interested reader (see Tables 6.3–6.7).

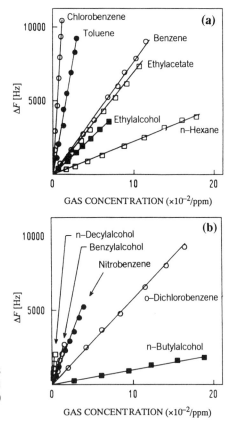

Figure 6.16. Linear response of the PQCMB as a function of concentration for several organic gases in agreement with Eq. (6.49) (Kurosawa et al., 1990).

Table 6.3. Interferences in the Assay of Ammonia Using an L-Glutamic Acid·HCl Coating[a]

Interference Gas	Triphenylamine Coating ΔF (Hz)	Trimethylamine·HCl Coating ΔF (Hz)
H_2S	14	82
SO_2	25	40
CO	24	8
NO_2	36	77
NH_3	26	-537[b]
CO_2	30	26
TMA[c]	41	134

Source: Guilbault (1984).
[a]The response of the PQCMB to the gaseous pollutants is at the concentration stated.
[b]Irreversible process.
[c]TMA = trimethylamine.

Table 6.4. Interferences in the Assay of Ammonia Using a Pyridoxime·HCl Coating[a]

Interference	Concentration (ppm)	ΔF (Hz)
NH_3	0001	1190
CO	1000	0033
HCl	0100	0043
NO_2	0100	0030
H_2S	1000	0025
SO_2	0100	0038
CO_2	1000	0040
TMA	0100	0020
Dry air	—	0000

Source: Guilbault (1984).
[a]The response of the PQCMB to some gaseous pollutants is at the concentrations stated.

Table 6.5. Interferences in the Assay of HCl Gas[a]

Interference	Concentration (ppm)	ΔF (Hz)
NH_3	0010	740
CO	1000	000
HCl	0100	020
NO_2	0100	025
H_2S	1000	059
CO_2	1000	036
Dry air	—	000

Source: Guilbault (1984).
[a]The response of the PQCMB to some gaseous pollutants is at the concentrations stated.

Table 6.6. Response to Interferences of a PQCMB with a Copper Chelate Coating

Interference[a]	Concentration (ppm)	ΔF (Hz)
Carbon monoxide	1250	0
Ammonia	1250	26
	250	2
Hydrogen sulfide	1250	18
	250	0
Sulfur dioxide	1250	12
	250	4
Hydrogen chloride	1250	40[b]
	250	15[b]
Benzene	1000	9
Toluene	500	13
Chloroform	500	49
	100	11

Source: Guilbault (1984).
[a]Frequency change, ΔF, to 15 ppm of DIMP (diisopropylmethyl phosphonate) = 591 Hz.
[b]Irreversible.

Table 6.7. Figures of Merit for a A–PQCMB

Detected Gas	Sensitivity (Hz/ppm)	Selectivity
m-Dichlorobenzene	0.141	0.170
1,1,2-Trichloroethane	0.023	0.170
2-Methyl-2-pentanol	0.058	0.620
Vapor	0.010	0.620
m-Dichlorobenzene	0.093	0.125
1,2-Dichloropropane	0.010	0.186
1,1,2-Trichloroethane	0.018	0.145

Source: Carey et al. (1987).

6.5. ARRAY OF PIEZOELECTRIC QUARTZ CRYSTAL MICROBALANCES

One of the main problems in solid state gas sensor technology is the absence of selectivity of some of these devices. In fact, selectivity is one of the main characteristics of gas sensors which, together with the sensitivity, defines the quality of the device. The selectivity is, of course, high in the case where interfering gases are molecularly very different, but it is low and sometimes nonexistent when the interfering species are of the same molecular families (hydrocarbons, alcohol, etc.). To avoid this problem, the active adsorptive coating employed with these devices as a catalyst must be totally selective for a single component. For example, palladium metal is very effective in the case of hydrogen detection owing to its properties (solubility of hydrogen); nevertheless, palladium can also adsorb other gases which contain hydrogen molecules such as H_2S, alcohols, and hydrocarbons. In the case where the sensor device is present in an environment with many interfering gases, the problem of selectivity is obvious and is difficult, if not impossible, to overcome with a single detector.

As with every solid state sensor, PQCMB is very often confronted with such problems due to multiconcentrations. In the middle of the 1980s some experimental groups around the world (Carey et al., 1987; Carey and Kowalski, 1988) had the idea to introduce the techniques of multicomponent analysis by using an array of piezoelectric quartz crystal microbalances (A–PQCMB). All the quartz substrates are coated with different partially selective coating materials. As Carey et al. (1987) pointed out, the advantage of an array device of this type is that the A–PQCMB response for each gas corresponds to a fingerprint response pattern with component identification analogous to spectroscopy. More simply one can say that in the presence of n gaseous components one needs n sensors in order to have a system of the same number of equations and unknowns. Thus, the general data obtained by the A–PQCMBs after multivariate calibration allows one to decouple the different interferences between the gases and to obtain an excellent idea as to the concentration of each gas in the studied environment. The A–PQCMB yields high selectivity measurements. Using the simple mathematics involved, and again taking the example of n equations and unknowns, we note that this system of n sensor signals and n gaseous components is easy to solve if the coefficients of the system can be easily decoupled, i.e., if they are independent. In the case of A–PQCMB, the term *independent coefficients* means that the detectable gases do not contain common molecules.

The quantitative analysis of multicomponent gas sample requires the use of a mathematical multivariate calibration. Two multivariant calibration techniques have been used by Carey et al. (1987): multiple linear regression (MLR) and partial least squares (PLS). In matrix notation MLR can be written

$$C(i, k) = R(i, j)\Xi(j, k) + E(i, k) \qquad (6.53)$$

where C is an $i \times k$ matrix of k analyte concentrations in each i sample; R is an $i \times j$ matrix of sensor responses to i samples of j sensors; Ξ is a $j \times k$ matrix of regression coefficients; and finally E is an $i \times k$ matrix of residuals. During the calibration process the matrices R_0 and C_0 present the measured responses and the calibration concentrations, respectively. Once these are determined, rearranging the above equations and solving for Ξ yields

$$\Xi = (R_0^T R_0)^{-1} R_0^T C_0 \qquad (6.54)$$

After estimation of Ξ, additional samples can be used and a quantitative analysis can be performed by substitution of the measured response vector into Eq. (6.54). One of the main problems with the use of MLR is the inversion of R due to a high degree of collinearity (mathematically, lack of independence between the columns; experimentally, gases with the same molecules). In fact, the significant difference that distinguishes PLS from MLR is that no matrix inversion is necessary and collinearity has no direct effect in the calibration model. Thus, once the PLS is established, a coefficient matrix can be calculated by using analogous regression coefficients found by MLR (Naes and Martens, 1985).

Figure 6.17 shows the experimental apparatus of the piezoelectric crystal sensor array (A–PQCMB), which consists of n piezoelectric crystals. The n crystals are coated with different adsorption films, and n oscillators are installed in order to excite the quartz crystals. Their output is connected to

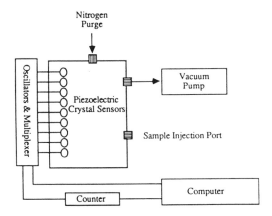

Figure 6.17. Instrumental diagram of piezoelectric crystal sensor array (Carey et al., 1987).

multifrequency meters and finally to a computer that collects the frequency change of each crystal. Once the data have been stored, the mathematical analysis can be performed.

Table 6.7 shows some important results obtained by a A–PQCMB system with an analysis using the foregoing mathematical model. In this table one can see the frequency sensitivity and selectivity of the device for six gases. The sensitivity has been defined as

$$\text{sensitivity} \equiv \frac{\text{a unique portion of the response pattern}}{\text{the analyte concentration of the response}} \qquad (6.55)$$

The selectivity presented in Table 6.7 has been defined as the ratio of the A–PQCMB response for one particular gas in the total mixture over the response in the presence of this particular pure gas:

$$\text{selectivity} \equiv \frac{X \text{ gas in the total mixture}}{\text{pure } X \text{ gas}} \qquad (6.56)$$

Thus, the selectivity ratio is always between 0 and 1. A close study of the above results has shown the superiority of the PLS model over the MLR model. According to the above, the average prediction error for PLS is approximately 10% whereas that for MLR is greater than 18%.

6.6. OTHER APPLICATIONS OF THE PQCMB

The detection of gases is not the only application of the PQCMB. In fact, during the last 40 years the PQCMB has been used in many areas of analytical

Table 6.8. Various Nongaseous Applications of the PQCMB

References	Application
Oberg and Logensjo (1959); Warner (1960, 1965); Behrndt and Love (1962)	Film thickness and deposition rate
Mieure (1968)	Electrogravimetric metal trace analysis
Olin and Sem (1971)	Mass concentration of suspended particles
Daley and Lundgren (1975)	Aerosol mass
King (1964); Karasek and Tiernay (1974)	Gas chromatography
Schulz and King (1973)	Liquid chromatography
King (1970)	Polymer research
Skarström (1960)	Detection in liquids
Haller and White (1963)	Study of polymerization of butadiene
Fischer and King (1967)	Study of surface oxidation
Pacy (1959)	Pressure
Roederer and Bastiaans (1983)	Microgravimetric immunoassay
van Ballegooijen (1984)	Temperature
Blackburn (1951)	Interferometric studies

chemistry. Several workers used the piezoelectric device in gas and liquid chromatography and as a high-resolution sorption device. The piezoelectric sensor has also found several other applications in polymer research, in metal oxidation studies, and in monitoring mass concentrations of suspended particles. In addition, it has provided a method for measuring film growth and thicknesses; Table 6.8 reviews various nongaseous applications of the piezoelectric crystal and cites the scientists who developed these techniques.

REFERENCES

Abe, S., and Hosoya, T. (1984). *Proc. World Hydrogen Energy Conf., 5th*, Toronto, *1984*, Vol. 4, p. 1893.

Bastiaans, C.J. (1988). In *Chemical Sensors* (T.E. Edmonds, Ed.), Chapter 14. Chapman & Hall, New York.

Behrndt, K.H., and Love, R.W. (1962). *Vacuum* **12**, 1.

Blackburn, R.K. (1951). U.S. Patent 2,536,025.

Bucur, R.V. (1981). *J. Catal.* **70**, 92.

Bucur, R.V., Mecea, V., and Flanagan, T.B. (1976a). *Surf. Sci.* **54**, 477.

Bucur, R.V., Mecea, V., and Indrea, E.I. (1976b). *J. Less-Common Met.* **49**, 147.

Carey, W.P., and Kowalski, B.R. (1988). *Anal. Chem.* **60**, 541.

Carey, W.P., Beebe, K.R., and Kowalski, B.R. (1987). *Anal. Chem.* **59**, 1529.

Cheney, J.L., and Homolya, J.B. (1975). *Anal. Lett.* **8**, 175.

Cheney, J.L., and Homolya, J.B. (1976). *Sci. Total Environ.* **5**, 69.

Cheney, J.L., and Norwood, T., and Homolya, J.B. (1972). *Anal. Lett.* **9**, 361.

Christen, M. (1983). *Sens. Actuators* **4**, 555.

Christofides, C., and Mandelis, A. (1989). *J. Appl. Phys.* **66**, 3986.

Christofides, C., and Mandelis, A. (1990). *J. Appl. Phys.* **68**, R1.

Cooper, J.B., Newbower, R.S., Sebok, D.A., and Meng, T.K. (1979). *Proc. Ann. Conf. Eng. Med. Biol.*, 32nd, Vol. 21, p. 15.

Cooper, J.B., Newbower, R.S., Sebok, D.A., and Meng, T.K. (1981). *IEEE Trans. Biomed. Eng.* **BME-28**, 459.

Czanderna, A.W., and Lu, C. (1984). In *Applications of Piezoelectric Quartz Crystal Microbalance* (C. Lu and A.W. Czanderna, Eds.), Vol. 7, Chapter 1. Elsevier, Amsterdam.

Daley, P.S., and Lundgren, D.A. (1975). *Am. Ind. Hyg. Assoc. J.* **36**, 518.

Daniels, F., and Alberty, R.A. (1966). *Physical Chemistry*, 3rd ed. Wiley, New York.

Deakin, M.R., and Byrd, H. (1989). *Anal. Chem.* **61**, 290.

DeGroot, S.R., and Mazur, P. (1969). *Non-Equilibrium Thermodynamics.* North-Holland Publ., Amsterdam.

Earp, R.B.W. (1966). Ph.D. Thesis, University of Alabama, Tuscaloosa.

Eer Nisse, E.P. (1967). *IEEE Trans. Sonics Ultrason.* **SU-14**, 59.

Everett, D.H., and Nordon, P. (1960). *Proc. R. Soc. London Ser.*, A **259**, 341.

Fischer, W.F., and King, W.H. (1967). *Anal. Chem.* **39**, 1265.

Frazier, G.A., and Glosser, R. (1979). *J. Phys. D* **12**, L113.

Frazier, G.A., and Glosser, R. (1980). *J. Less-Common Met.* **74**, 89.

Frechette, M.W., and Fasching, J.L. (1973). *Environ. Sci. Technol.* **7**, 1135.

Guilbault, G.G. (1984). In *Applications of Piezoelectric Quartz Crystal Microbalance* (C. Lu and A.W. Czanderna, eds.), Vol. 7, Chapter 8, p. 251. Elsevier, Amsterdam.

Guilbault, G.G., and Ho, M. (1982). *Anal. Chem.* **54**. 1998.

Guilbauld, G.G., and Lopez-Roman, A. (1975). *Environ. Lett.* **2**, 35.

Guilbault, G.G., and Tomita, Y. (1980). *Anal. Chem.* **52**, 1484.

Guilbault, G.G., Ho, M., and Rietz, B. (1980). *Anal. Chem.* **52**, 1489.

Guilbault, G.G., Affolter, J., Tomita, Y., and Kolesar, E.S. (1981). *Anal. Chem.* **53**, 2057.

Guilbault, G.G., Tomita, Y., and Kolesar, E.S. (1981/82). *Sens. Actuators* **2**, 43.

Haller, I., and White, P. (1963). *Rev. Sci. Instrum.* **34**, 677.

Hlavay, J., and Guilbault, G.G. (1978a). *Anal. Chem.* **50**, 965.

Hlavay, J., and Guilbault, G.G. (1978b). *Anal. Chem.* **50**, 1044.

Ho, M.H., Guilbault, G.G., and Rietz, B. (1980). *Anal. Chem.* **52**, 1489.

Janghorbani, M., and Freund, H. (1973). *Anal. Chem.* **45**, 325.

Karasek, F.W., and Tiernay, J.M. (1974). *J. Chromatogr.* **89**, 31.

Karmarkar, K.H., and Guilbault, G.G. (1974). *Anal. Chim. Acta* **71**, 419.

Karmarkar, K.H., and Guilbault, G.G. (1975). *Anal. Chim. Acta* **75**, 111.

Karmarkar, K.H., Webber, L.M., and Guilbault, G.G. (1975). *Environ. Lett.* **8**, 345.

Karmarkar, K.H., Webber, Ł.M., and Guilbault, G.G. (1976). *Anal. Chim. Acta* **81**, 265.

Kasemo, B., and Tornqvist, E. (1980). *Phys. Rev. Lett.* **44**, 1555.

Kindlund, A., and Lundström, I. (1982/83). *Sens. Actuators* **3**, 63.

King, W.H. (1964). *Anal. Chem.* **36**, 1735.

King, W.H. (1969). *Res. Dev.* **20**, 28.

King, W.H. (1970). U.S. Patent 3,164,004.

King, W.H. (1981). *Vac. Microbalance Tech.* **8**, 25.

Kristoff, J., and Guilbault, G.G. (1983). *Anal. Chim. Acta.* **149**, 337.

Kurosawa, S., Kamo, N., Matsui, D., and Kobatake, Y. (1990). *Anal. Chem.* **62**, 353.

Landau, L.D., and Lifschitz, E.M. (1959). *Theory of Elasticity.* Pergamon, Oxford.

Lasher, J.R. (1937). *Proc. R. Soc. London, Ser. A* **161**, 525.

Lee, C.W., Fung, Y.S., and Fung, K.W. (1982). *Anal. Chim. Acta* **135**, 277.

Lewis, F.A. (1967). *The Palladium–Hydrogen System.* Academic Press, New York.

Llopis, J. (1968). *J. Catal. Rev.* **2**, 161.

Lopez-Roman, A., and Guilbault, G.G. (1972). *Anal. Lett.* **5**, 225.

Lovinger, A.J. (1983). *Science* **220**, 4602.

Lu, C. (1984). In *Applications of Piezoelectric Quartz Crystal Microbalance* (C. Lu and A.W. Czanderna, Eds.), Chapter 2. Elsevier, Amsterdam.

Lu, C., and Lewis, O. (1972). *J. Appl. Phys.* **43**, 4385.

Lundström, K.I., Armgarth, M., and Petersson, L.-G. (1989). *CRC Crit. Rev. Solid State Mater. Sci.* **15**, 201.

Maarsen, F.W., Smit, M.C., and Matze, J. (1957). *Recveil* **76**, 713.

Mason, W.P. (1950). *Piezoelectric Crystals and Their Application to Ultrasonics.* Van Nostrand–Reinhold, Princeton, NJ.

Mecea, V., and Bucur, R.V. (1980). *J. Vac. Sci. Technol.* **17**, 182.

Mecea, V., and Ghete, P. (1984). *Int. J. Hydrogen Energy* **9**, 861.

Mieure, J.P. (1968). Ph.D. Thesis, Texas A&M Univ., College Station, TX.

Miller, J.G., and Bolef, D.I. (1968a). *J. Appl. Phys.* **39**, 4589.

Miller, J.G., and Bolef, D.I. (1968b). *J. Appl. Phys.* **39**, 5815.

Naes, T., and Martens, H. (1985). *Commun. Stat.–Simula Computa* **14**, 5450.

NASA Tech. Brief (1968). **MFS-23357**.

Oberg, P., and Lingensjo, G. (1959). *Rev. Sci. Instrum.* **30**, 1053.

Olin, J.G., and Sem. G.J. (1971). *Atmos. Environ.* **5**, 653.

Pacy, D.J. (1959). *Vacuum* **9**, 261.

Ponec, V., Knor, Z., and Cerny, S. (1969). *Discuss. Faraday Soc.* **41**, 149.

Randin, J.P., and Zulling, F. (1987). *Sens. Actuators* **11**, 319.

Roederer, J.E., and Bastiaans, G.J. (1983). *Anal. Chem.* **55**, 2333.

Sauerbrey, G.Z. (1959). *Z. Phys.* **155**, 206.

Sauerbrey, G.Z. (1964). *Z. Phys.* **178**, 547.

Scheide, E.P., and Guilbault, G.G. (1972). *Anal. Chem.* **44**, 1764.

Scheide, E.P., and Taylor, J.K. (1974). *Environ. Sci. Technol.* **8**, 1097.

Scheide, E.P., and Taylor, J.K. (1975). *Am. Ind. Hyg. Assoc. J.* **36**, 897.

Schulz, W.W., and King, W.H. (1973). *J. Chromatogr.* **11**, 343.

Shackleford, W.H., and Guilbault, G.G. (1974). *Anal. Chim. Acta* **73**, 383.

Sermon, P.A. (1972). *J. Catal.* **24**, 460.

Simons, J.W., and Flanagan, T.B. (1965a). *Can. J. Chem.* **43**, 1665.

Simons, J.W., and Flanagan, T.B. (1965b). *J. Phys. Chem.* **69**, 3773.

Skarström, C.W. (1960). U.S. Patent 2, 294, 627.

Smith, A.C., Janak, J.F., and Adler, R.B. (1967). *Electronic Conduction in Solids* McGraw-Hill, New York.

Steele, M.C., Hile, J.W., and MacIver, B.A. (1986). *J. Appl. Phys.* **47**, 2537.

Stockbridge, C.D. (1966). In *Vaccum Microbalance Techniques* (K.H. Behrndt, Ed.), Vol. 5, p. 193. Plenum, New York.

Tomita, Y., Ho, M.H., and Guilbault, G.G. (1979). *Anal. Chem.* **51**, 1475.

van Ballegooijen, E.C. (1984). In *Applications of Piezoelectric Quartz Crystal Micro-balance* (C. Lu and A.W. Czanderna, Eds.), Chapter 5. Elsevier, Amsterdam.

Vannice, M.A., Benson, J.E., and Boudart, M. (1970). *J. Catal.* **16**, 348.

Warner, A.W. (1960). *Bell Syst. Tech. J.* **39**, 1193.

Warner, A.W. (1965). *IEEE Trans. Sonics Ultrason.* **SU-12**, 2.

Warner, A.W., and Stockbridge, C.D. (1962). In *Vacuum Microbalance Techniques* (R.F. Walker, Ed.), Vol. 2, p. 71. Plenum, New York.

Webber, L.M., and Guilbault, G.G. (1976). *Anal. Chem.* **48**, 2344.

Webber, L.M., Karmarkar, K.H., and Guilbault, G.G. (1978a). *Anal. Chim. Acta* **50**, 29.

Webber, L.M., Karmarkar, K.H., and Guilbault, G.G. (1978b). *Anal. Chem.* **97**, 29.

Zemel, J.N. (1985). In *Solid State Chemical Sensors* (J. Janata and R.J. Huber, Eds.), Chapter 4. Academic Press, New York.

Zemel, J.N. (1990). *Rev. Sci. Instrum.* **61**, 1579.

CHAPTER
7

SURFACE ACOUSTIC WAVE SENSORS

7.1. INTRODUCTION AND HISTORICAL PERSPECTIVE

The mathematical basis of the phenomenon of surface acoustic waves (SAW) was first published last century by Lord Rayleigh (1885). At that time, SAW drew the attention of the scientific community and especially that of geologists because the acoustic energy released by earthquakes moves as a surface acoustic wave on the earth's crust. This natural type of waves is known today as Rayleigh waves. The main characteristic of SAW is that the acoustic field and particle displacement occur mainly on the surface of the solid. Almost a century had to pass before SAW found their first application in electronics. In the mid-1960s, White and Voltmer (1965) developed an interdigital transducer (IDT) that allowed the generation of Rayleigh surface waves in piezoelectric solids and then the application of SAW to radio-frequency (rf) signals. The main reason for the choice of Rayleigh waves among other kinds of SAW (such as Love and Stonely waves) is that they are generated quite readily in a variety of piezoelectric materials using an IDT.

During recent decades, SAW techniques have emerged rapidly in sensor technology owing to advances in our understanding of the mechanism of interaction between measurant and acoustic propagation in the piezoelectric substrates such as YZ-$LiNbO_3$, YZ- and ST-quartz, ZY-$LiTaO_3$, and AlN/Al_2O_3 (Morgan, 1973). This interaction of the SAW with the surrounding media includes various linear and nonlinear properties of the piezoelectric medium, such as mass density, elastic stiffness, and electric and dielectric properties.

Moreover, the development of microelectronic silicon technology led rapidly to the fabrication of SAW sensors by use of the techniques of thin-film deposition, photolithography, as well as the technology of planar integrated circuits. This opened the way toward the miniaturization of such devices and considerable reduction of the cost (Verona, 1988). The application of SAW to signal processing remains of great importance to date (Campbell, 1989).

In 1976 a special IEEE issue (Proceedings of the IEEE, 1976) on surface acoustic wave devices and applications was published, and the importance of these structures for future applications became obvious. SAW devices have been employed largely in signal processing (Special Issue on Microwave

Acoustics, 1969; Matthews, 1977; Oliner, 1978) and in both analog and digital filters and resonators. Since then, several extensive reviews on SAW device sensors have been published (Verona, 1988; D'Amico and Verona, 1988, 1989; Zemel, 1990; Christofides and Mandelis, 1990).

Section 7.2 presents the theoretical discussion. Not much fundamental theoretical development is given there, since the piezoelectric theory has already been presented in Chapter 6. Nevertheless, the section contains essential details of SAW effects on piezoelectric substrates. Some other properties of SAW materials are discussed as well.

Section 7.3 deals with the experimental apparatus and delineates the principle of experimental SAW setups. Some variations in apparatus are indicated.

Section 7.4 provides an extensive overview of experimental results obtained since the application of SAW devices to gas monitoring. Some tables show the types of coatings that have been used for various gas detection schemes. Moreover, a comparison of a bulk wave device (PQCMB; see Chapter 6) and a surface wave device for the detection of H_2 and SO_2 is presented. Several factors that influence the SAW response, such as types of substrates and chemical coatings, thickness of coating, temperature, and gaseous interferences, are considered in detail.

Section 7.5 presents an analytical comparison of the main piezoelectric devices established by Wenzel and White (1989a, b), while Section 7.6 describes a SAW array system with pattern recognition capabilities.

Finally, Section 7.7 presents other applications of surface acoustic devices. A table summarizes much information concerning the capabilities of SAW device sensors.

7.2. THEORY OF SAW DEVICES

In recent years three main types of SAW device structure have been used for gas as well as for liquid detection. These main surface acoustic devices are conventional Rayleigh surface wave devices, plate-mode structures, and transverse wave elements.

7.2.1. Rayleigh SAW Sensors

The principle of the SAW detectors is simple (Wohltjen, 1984; Wohltjen et al., 1985): when gas molecules are absorbed in a thin chemically coated layer, chosen according to the desired gas selectivity, they perturb the properties of propagating SAW in the piezoelectric substrate. In other words, the energy of SAW is localized within one or two acoustic wavelengths from the surface.

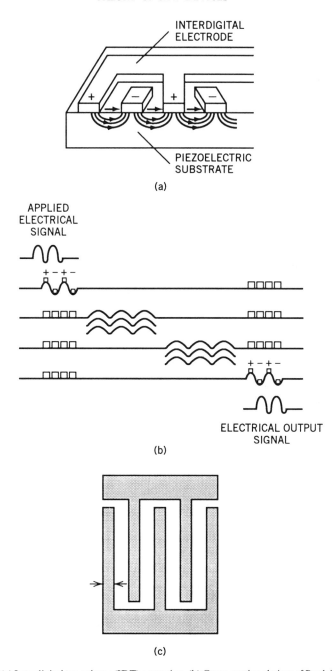

Figure 7.1. (a) Interdigital transducer (IDT) operation. (b) Cross-sectional view of Rayleigh wave operation. (c) SAW interdigital transducer design (Wohltjen and Dessy, 1979).

This property enables the SAW to interact easily and strongly with the medium adjacent to the surface. Wave–medium interaction is the main necessary (but not sufficient) property that can lead to a gas detector. The second question for device applications is to choose among various kinds of SAW, such as Love, Stonely, and Rayleigh waves. The primary reason for focusing on Rayleigh waves is that they are generated quite easily in many piezoelectric substrates using an interdigital transducer electrode (Wohltjen, 1984). Thus, the generation of the SAW by IDT is followed by the propagation of the wave through a chemically selective coated layer. This wave–matter interaction causes a perturbation of the characteristics of the Rayleigh surface waves such as amplitude, phase, and velocity. Figure 7.1(a) is a schematic presentation of an IDT operation, and Fig. 7.1(b) presents a cross-sectional view of Rayleigh wave propagation (Wohltjen and Dessy, 1979). In Fig. 7.1(c) one can see in detail the intertwined finger (interdigital) arrangement of an IDT used for the generation as well as for the detection of surface waves. The measurement of changes in the SAW characteristics is an indicator of the presence of a gas on the coating surface layer of the device. In essence, there are three different modes of detection for monitoring gases: (a) amplitude changes in the waveguiding layer take place because of the coupling of energy from the surface into the adjacent layer of gaseous matter; (b) the velocity change of SAW propagation introduces phase shifts in the wave; and (c) the resonant frequency, F, of the SAW device oscillator changes because of contact between the absorbed matter (gas) and the coating layer. In fact, there exists a relation between the fractional changes in the oscillation frequency, F, the acoustic velocity, U, and the phase delay Φ (D'Amico et al., 1982):

$$\frac{\Delta F}{F} = \frac{\Delta \Phi}{\Phi} = -\frac{\Delta U}{U} \tag{7.1}$$

7.2.1.1. SAW Detector Amplitude Perturbations

A schematic representation of the SAW gas sensor based on amplitude change is shown in Fig. 7.2 (D'Amico et al., 1982/83). The operating characteristics of the detector are described in Table 7.1. This SAW detector has been used by D'Amico et al. (1982) for hydrogen detection. Surface acoustic waves were generated by an input transducer T and collected by two output transducers, T_R and T_S, which were located at opposite sides of the YZ-LiNbO$_3$ substrate. The hydrogen Pd–SAW detector was fabricated on YZ-LiNbO$_3$ piezoelectric substrate. One of the propagation paths (l_S) was coated with a Pd thin-film layer that acted as the selective layer of the device; on the other hand, the other path (l_R) was uncoated and it was used as a reference. The voltage

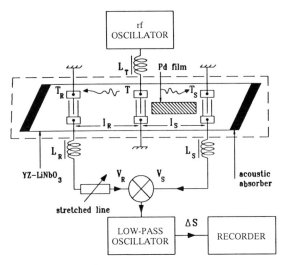

Figure 7.2. Schematic diagram of the SAW detector setup based on amplitude change (D'Amico et al., 1982/83). See the text for description and details.

Table 7.1. Operating Characteristics of the Pd–SAW Hydrogen Detector

Pd–SAW Sensor Characteristics	Values
Operating frequency, F (MHz)	75–76
Reference length, L_s (mm)	15.5
Palladium length, L_R (mm)	13.5
Palladium thickness, L (Å)	3000
Number of IDT finger pairs	5

Source: D'Amico et al. (1982/83).

outputs, V_S and V_R, through inductances L_S and L_R, were connected to a double-balanced mixer and then the output differential voltage, ΔS, was filtered out and recorded as (D'Amico et al., 1982/83):

$$\Delta S = V_R - V_S = V_M \sin(\Phi_R - \Phi_S) = V_M \sin(\Delta\Phi) \qquad (7.2)$$

where V_M is the maximum output voltage (when $\Delta\Phi = K_n\pi/2$; here K_n is a positive integer). From the wave propagation delay one may write the phase shift as (D'Amico et al., 1982/83)

$$\Delta\Phi = 2\pi F\left(\frac{l_R}{U_R} - \frac{l_S}{U_S}\right) + \Phi_E \qquad (7.3)$$

where U_R and U_S are the velocities of the acoustic waves in the coated and noncoated surfaces, respectively, and Φ_E is the electric phase shift introduced by a stretched line used to adjust the initial phase difference $\Delta\Phi$ in order to minimize the differential output signal, ΔS ($\Delta S \to 0$). When adsorption–absorption (or desorption) of hydrogen takes place on the Pd surface, it changes the density and the elastic properties of the film, which causes a shift on the SAW propagation velocity. As a consequence a voltage change is detected by the device as a function of hydrogen concentration C_0 (D'Amico et al., 1982/83):

$$\delta S = \frac{\partial V}{\partial C_0} = 2\pi V_M \frac{F l_S}{U_S^2}\left(\frac{\partial U_S}{\partial C_0}\right)\cos(\Delta\Phi) \qquad (7.4)$$

As the argument of the cosine function was minimized ($\Delta\Phi \approx 0$) at the beginning of each experiment by the initial phase zeroing procedure, Eq. (7.4) can be expressed as

$$\delta S = 2\pi V_M \frac{F l_S}{U_S^2}\left(\frac{\partial U_S}{\partial C_0}\right) \qquad (7.5)$$

It can be seen from Eq. (7.5) that in the case of a SAW amplitude detector for a given substrate material, the sensitivity depends on the $F \times l_S$ product; thus it is clear that for a fixed sensitivity the smaller the device to be fabricated, the higher its operating frequency, F, must be.

7.2.1.2. SAW Detector Frequency (or Velocity) Perturbations

The majority of SAW devices are based on phase velocity changes for monitoring the measurand (Verona, 1988). In fact, two different methods are generally used: (a) the phase shift, and (b) the frequency shift detection technique.

Wohltjen's group has shown that the shift frequency, ΔF, of a SAW gas detector can be expressed by the following relationship (Wohltjen, 1984; Wohltjen et al., 1985):

$$\Delta F = [(k_1 + k_2)F^2 L\rho] - \left(k_2 F^2 L \frac{4\mu}{U_R^2}\sigma\right) \qquad (7.6)$$

where k_1 and k_2 are material constants for the SAW substrate [see Table 7.2 (Auld, 1973)]; L is the thickness of the Pd film; ρ is the film density ($m = \rho \times L$: mass per unit area); and σ is given by the relation

$$\sigma = \frac{\lambda + \mu}{\lambda + 2\mu} \qquad (7.7)$$

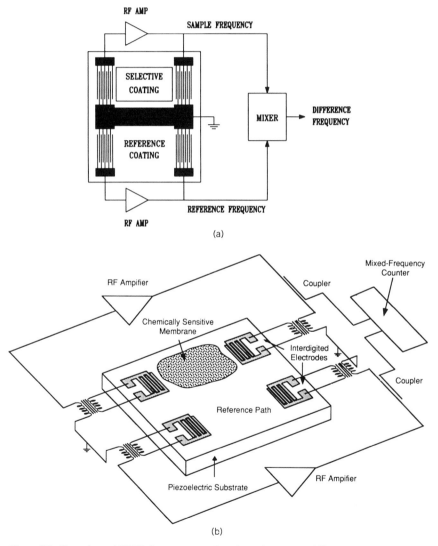

Figure 7.3. Top view of SAW detector setup based on frequency shift measurements: (a) by Wohltjen et al. (1985) and Ballantine et al. (1986); (b) by Zemel (1990).

Table 7.2. Material Constants for Selected SAW Substrates

Substrate	V_R (m/s)	k_1 (m²s/kg) × 10⁻⁸	k_2 (m²s/kg) × 10⁻⁸
Y cut X propagation quartz	3159.3	-9.33	-4.16
Y cut Z propagation LiNbO₃	3487.7	-3.77	-1.73
Z cut X or Y propagation CdS	1702.2	-8.33	-2.67
Z cut X or Y ZnO	2639.4	-5.47	-2.06
Z cut X propagation Si	4921.2	-9.53	-6.33

Source: Auld (1973).

where λ and μ are the film Lamé constants. In Eq. (7.6) the dependence of the sensor response on the gas concentration is pronounced through the variation of m. In the case where the chemically selective layer is an elastomeric organic polymer, Eq. (7.6) may be simplified to (Wohltjen, 1984):

$$\Delta F \approx (k_1 + k_2)F^2 \frac{\Delta m}{A_s} \tag{7.8}$$

where Δm is the absorbed gas mass and A_s is the surface of the coated layer. In Eq. (7.8) one can point out the formal analogy to Eq. (6.47) valid for piezoelectric quartz crystal microbalance detectors. Figure 7.3(a) shows an experimental setup by Wohltjen et al. (1985) for vapor detection using frequency shift output. The same experimental setup has been used by Ballantine et al. (1986). The SAW device was coated with a polymer layer and was exposed to small concentrations of dimethyl methylphosphonate (2 ppm) in a dry-air carrier stream at two operating frequencies (31 and 112 MHz). The dual configuration improved the detector stability against temperature and pressure changes. Figure 7.3(b) presents another excellent design of the SAW where one can recognize more details of the electronic circuit. Table 7.2 shows typical material constants for selected SAW substrates.

7.2.2. Plate-Mode Structures

White et al. (1987) have investigated the properties of thin silicon plates that have been excited into Lamb wave oscillations. According to these authors the fundamental condition for the existence of such Lamb waves is that the thickness of membranes be small compared to the wavelength of the acoustic excitation. One can distinguish two types of Lamb waves: (a) symmetric Lamb waves (SLW) and (b) antisymmetric Lamb waves (ALW). Figure 7.4

Longitudinal Motion

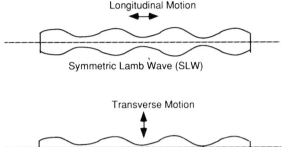

Symmetric Lamb Wave (SLW)

Transverse Motion

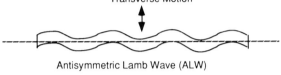

Antisymmetric Lamb Wave (ALW)

Figure 7.4. Symmetric and antisymmetric Lamb wave modes in a thin plate (Zemel, 1990).

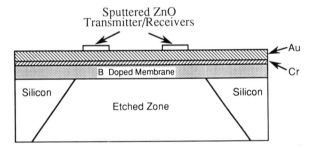

Figure 7.5. Cross section of the Lamb wave antisymmetric structure (Wenzel and White, 1988a).

(Zemel, 1990) shows the SLW and ALW modes in a thin plate. Figure 7.5 is a cross section of the ALW structure after Wenzel and White (1988a, b). In fact the Lamb wave can be generated by an interdigited electrode pair on the ZnO and received by a similar electronic structure. Wenzel and White (1989a) proved that the phase velocity, V_p, for the plate-mode sensor is given by the relation

$$V_p = \frac{2\pi}{\lambda} \left(\frac{B}{m + \Delta m} \right)^{1/2} \tag{7.9}$$

where λ is the flexural-mode wavelength; m, the unloaded mass per unit area; and B, the effective plate stiffness. Equation (7.9) is valid only for thin plates with thickness $d \ll \lambda$. Finally $\Delta m = \rho_f \delta$, where ρ_f is the fluid density and δ is the skin depth of the evanescent disturbance in the fluid and depends on the sound velocity of the fluid, which has to be lower than the phase velocity.

In fact, Wenzel and White (1988a, b) showed that the ALW are sensitive to the mass loading of the films when they are in contact with a liquid where V_p of the ALW is smaller than the bulk sound velocity of the fluid V_f. In practical terms this is a very important result because it leads to the conclusion that ALW chemical sensing could become effective for liquid chemistries. The Lamb wave structure can operate even when the device is immersed in fluids. Before closing this subsection we should mention that Wenzel and White (1989a, b) proved that generation of plate waves with electrostriction and an optical detector scheme is possible. This principle has been used for detection of organic vapors in the ppb range.

7.2.3. Transverse Wave Elements

The application of surface transverse waves (STW) to the detection of gases has been established during the last few years (Anisimkin et al., 1989). STW seem attractive compared to Rayleigh waves owing to the possibility of varying their penetration depth below the surface; as was pointed out in Section 7.2.1, the Rayleigh waves propagate only in a very thin layer on the surface of the substrate. This helps increase the sensitivity of the device by simply increasing the acoustic power density in the coated chemical layer. As in the case of SAW sensors, STWs can be induced and detected by means of IDT. It is important to note that the elastic constants that enter into the propagation of STWs are different from those of SAWs. Figure 7.6 shows the direction of the propagation of both STW and SAW on the same ST-cut quartz substrate. The experimental results of Anisimkin et al. (1989) pertaining to STW detection are discussed later in this chapter.

Study of the interaction of transverse (shear mode) waves with liquids has progressed apace in recent years. Thompson et al. (1987) have used these

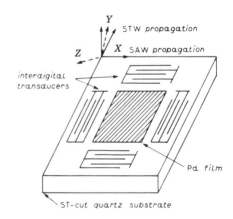

Figure 7.6. Schematic of the STW and SAW hydrogen sensor implemented on the same ST-cut quartz substrate (Anisimkin et al., 1989).

wave modes to monitor biological interactions, and Ricco and Martin (1988) and Ricco et al. (1988) have studied the interactions of inorganic materials at the liquid–solid interface. That work, however, is beyond the scope of this book.

7.3. SAW EXPERIMENTAL INSTRUMENTATION

Most SAW sensors have been implemented on piezoelectric substrates such as $LiNbO_3$ crystals or quartz. Several workers have also used thermally oxidized silicon substrates with piezoelectric overlays of ZnO (Vellekoop et al., 1987). Rayleigh and Lamb acoustic propagation modes have been used for the design of SAW devices. Unfortunately, few papers in the field deal with the design procedure of such devices or with their optimization; thus a technical review is not yet possible owing to the paucity of information in the literature.

The most common structure used for the implementation of SAW is a delay-line SAW oscillator (see Fig. 7.7). Dual differential structures are also shown in Fig. 7.3. This technique is very common in SAW detection because it is more sensitive and compensating for factors such as changes in environ-

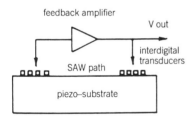

Figure 7.7. Schematic of a single delay-line SAW oscillator (D'Amico and Verona, 1989).

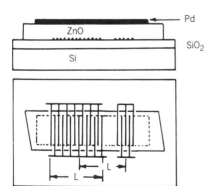

Figure 7.8. Schematic of a SAW H_2 sensor with interdigital electrodes between ZnO and SiO_2 layers (D'Amico and Verona, 1989).

Table 7.3. Type of Gases Detected by SAW Devices

References	Detection	Substrate	Coating
D'Amico et al. (1982)	Hydrogen	YZ-LiNbO$_3$	Palladium
D'Amico et al. (1986)	Hydrogen	STX-SiO$_2$	Palladium
Caliendo et al. (1988)	Hydrogen	ZnO/SiO$_2$/Si	Palladium
Arya et al. (1988)	Hydrogen	ZnO-coating glass	Palladium
Bryant et al. (1981)	SO$_2$	YZ-LiNbO$_3$	Triethanolamine
Bryant et al. (1983)	SO$_2$	YZ-LiNbO$_3$	Triethanolamine
Nieuwenhuizen et al. (1989)	SO$_2$	STX-SiO$_2$	PC
D'Amico and Verona (1989)	NO$_2$	YZ-LiNbO$_3$	Lead phthalocyanine
Barensz et al. (1985)	NO$_2$	STX-SiO$_2$	PC (phthalocyanine)
Ricco et al. (1985)	NO$_2$	YX-LiNbO$_3$	PC(Pd)
Venema et al. (1986)	NO$_2$	STX-SiO$_2$	PC
Venema et al. (1987)	NO$_2$	STX-SiO$_2$	PC
Nieuwenhuizen et al. (1989)	NO$_2$	STX-SiO$_2$	PC
Heckl et al. (1990)	Derivatives of nitrobenzene	?	APTES with silinol groups
Wenzel and White (1989a, b)	Toluene	ZnO/Al/Si$_x$N$_y$	Poly(dimethylsiloxane)
Joshi and Brace (1985)	Humidity	YX-LiNbO$_3$	Polyimide

Reference	Organic vapors	Si/SiO$_2$/ZnO	Polymers
Zellers et al. (1987)	Organic vapors	Si/SiO$_2$/ZnO	Polymers
Chuang and White (1981)	Vapors	ZnO–Si	Polymers
Wohltjen et al. (1987)	Vapors	STX-SiO$_2$	Polymers
Martin et al. (1987a, b)	Vapors	STX-SiO$_2$	?
Zellers et al. (1990)	Styrene Vapor	?	Trans-PtCl$_2$ Ethylene
Bryant et al. (1983)	H$_2$S	YZ-LiNbO$_3$	WO$_3$
Bryant et al. (1983)	H$_2$S	YZ-LiNbO$_3$	TEA
Vetelino et al. (1986)	H$_2$S	YZ-LiNbO$_3$	WO$_3$
Vetelino et al. (1987)	H$_2$S	YZ-LiNbO$_3$	WO$_3$
Nieuwenhuizen et al. (1989)	CO	STX-SiO$_2$	PC
Nieuwenhuizen and Nederlof (1990)	CO$_2$	?	Poly(ethyleneimine)
Nieuwenhuizen et al. (1989)	H$_2$O	STX-SiO$_2$	PC
Nieuwenhuizen et al. (1989)	CH$_4$	STX-SiO$_2$	PC
D'Amico et al. (1987)	NH$_3$	STX-SiO$_2$	Platinum
Arya et al. (1988)	NH$_3$	ZnO–glass	Palladium
Brace et al. (1988)	Water	YZ-LiNbO$_3$	Hygroscopic

mental conditions (temperature and pressure variations). In the case of gas chemical sensors use of SAW differential structures introduces some problems because the two structures cannot be completely identical (D'Amico and Verona, 1989).

Finally Fig. 7.8 shows an example of a structure representing a delay-line suitable for SAW obtained by implementing ZnO-on-Si microelectronic technology (Vellekoop et al., 1987; Caliendo et al., 1988). We note that the interdigital electrodes are located at the interface between the SiO_2 and ZnO layers. Owing to this design they can be fabricated during the same step as electrical connections of the integrated amplifier.

7.4. SAW FOR GAS DETECTION: EXPERIMENTAL RESULTS

In recent years, considerable research effort has been directed toward the development of SAW gas sensors. Several workers have used the SAW device for the detection of various gases. Table 7.3 cites references with the type of detected gas, as well as the chemical coating used.

7.4.1. Hydrogen Detection

While Wohltjen and Dessy's (1979) work led to the development of SAW gas sensors, it was D'Amico et al. (1982) who first demonstrated that a SAW device with an appropriate chemically sensitive coating (Pd) could be employed as hydrogen detector. In this subsection, we shall describe the gas sensor SAW device and mainly the work of D'Amico et al. (1982, 1982/83, 1986), who developed the capability of the Pd–SAW device as a hydrogen detector. The studies of these authors were performed under atmospheric pressure and room-temperature conditions. They have used the three conventional detection modes, amplitude, phase, and frequency, for monitoring hydrogen gas. A system that allowed testing of the response of the SAW device in a flowing H_2–N_2 mixture was used. Their system consisted of three subsystems: a gas control (three flow meters, input and output valves); a test cell containing the Pd-coated surface acoustic wave detector (Pd–SAW); and an electronic shift amplitude analysis instrumentation [rf oscillator, law-pass filter, and recorder (see Fig. 7.2)].

The device was exposed to 0.1% H_2 in N_2 (flow rate $= 810\,mL/min$), and the Pd–SAW response was monitored as a function of time. Typical experimental results are shown in Fig. 7.9. We note that it takes more than 20 min for the signal amplitude to reach saturation (δS_s); R_s denotes the saturation time. In Fig. 7.9 it is seen that a few minutes after the introduction of hydrogen

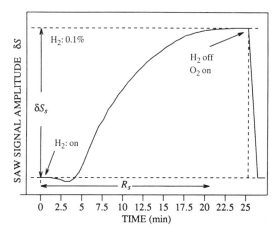

Figure 7.9. Time response of the Pd–SAW hydrogen detector during the absorption and desorption process (H_2 flow rate: 828 mL/min; O_2 flow rate: 150 mL/min; T = 20 °C) (D'Amico et al., 1982). The terms δS_s and R_s are defined in the text.

into the test cell the signal amplitude *decreases* as a function of time. The phenomenon of a negative amplitude shift upon H_2 exposure is surprising. No attempt by the authors to explain the decrease was made; however, it is possible that this phenomenon could be related to the anomalous behavior of the bulk piezoelectric quartz crystal (Pd–PQCMB) detector described in Chapter 6, i.e., the possibility that surface-absorbed gases other than H_2 may interfere with SAW operation.

Figure 7.10 shows the variation of saturation response amplitude, δS_s, as a function of gas phase hydrogen concentration in $N_2 + H_2$ fluxes in the range of 50 to 10^4 ppm, i.e., in the 0.005–1% range of H_2 in N_2. The numbers in parentheses in the figure show the response time (in minutes) as a function of H_2 concentration. The device was fabricated on a YZ-LiNbO$_3$ substrate and was operating at 75 MHz. The length of propagation through the chemical and reference layers was 15.5 mm. The Pd film thickness deposited on one of the substrates was 3000 Å with a length of 13.5 mm. Contrary to the bulk wave piezoelectric detector (PQCMB; see Chapter 6), the Pd–SAW sensor response seems to be influenced by the H_2 concentration even at very low concentrations. However, this comparison is only qualitative, since the thickness of the Pd chemical layer of the SAW device was only 3000 Å whereas it was 4300 Å in the silicon-based device.

Caliendo et al. (1988) have also presented some results concerning the detection of hydrogen by using a SAW sensor on silicon substrate. Fig. 7.11 shows the variation of $\Delta F/F$ vs. [H_2] in pure nitrogen. The sensor's palladium

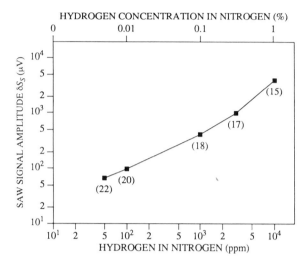

Figure 7.10. Variation of the saturation SAW response δS_s, as a function of hydrogen concentration at 20 °C. Numbers in parentheses: response time at shown hydrogen concentration in minutes. Rise and fall time values were measured at constant flow rates equal to 810 mL/min $(N_2 + H_2)$ and 150 mL/min (O_2) respectively (D'Amico et al., 1982).

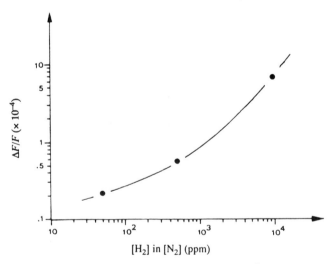

Figure 7.11. Fractional frequency change as measured on the 4300 Å Pd sensor upon exposure to various hydrogen concentrations in nitrogen (Caliendo et al., 1988).

layer was $4300\,\text{Å}$ thick. By comparing the fractional frequency change of Figs. 7.10 and 7.11 at $L = 4300\,\text{Å}$ for $[H_2] = 100\,\text{ppm}$ it is easy to see that the silicon-based device is almost 1 order of magnitude more sensitive than the piezoelectric device.

Anisimkin et al. (1989) used surface acoustic transverse waves for the detection of hydrogen gas in $H_2 + N_2$ gas mixtures. In the saturation region of hydrogen detection, oxygen is introduced in the cell in order to absorb the hydrogen from the chemically coated layer. This forces the device to reach its preexposure signal level. The experimental setup used is presented in Fig. 7.12. Table 7.4 compares results obtained by the aforementioned

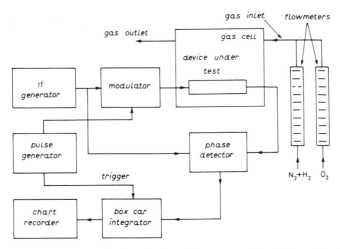

Figure 7.12. Experimental setup for investigation of the response of the device used by Anisimkin et al. (1989) to detect absorption and desorption of hydrogen.

Table 7.4. Fractional Velocity Change for SAW and STW Hydrogen Sensors upon Exposure to a Mixture of 1% $H_2 + N_2$, for Two Different Values of the Parameter h/λ^a

Acoustic Mode	h/λ $(\times 10^{-3})$	$\Delta v/v$ $(\times 10^{-6})$
SAW	1.6	50
STW	1.6	50
SAW	7.8	200
STW	7.8	130

Source: Anisimkin et al. (1989).

^{a}Here h/λ is the film thickness to wavelength ratio.

authors using SAW and STW devices. The main conclusion to be drawn from Table 7.4 is that the sensitivity of the STW device increases more rapidly than that of the SAW sensor as a function the h/λ parameter. It is also interesting to note that the sensitivity of the device reported by Anisimkin et al. (1989) has been found to be as much as 2 orders of magnitude higher than the Pd–LiNbO$_3$ reported earlier by the same group (D'Amico et al., 1982).

D'Amico et al. (1982) have also studied the influence of the Pd thickness on the sensitivity of the Pd–SAW detector, as well as the influence of flow rate on its response time R_s. For this investigation the above authors have explored the velocity change, ΔU, of the SAW. The relation between shift frequency and velocity change is

$$\frac{\Delta F}{F} = g\frac{\Delta U}{U} \tag{7.10}$$

where $g = L_R/L_S$ is a geometrical factor. The net increase in mass due to the H$_2$–Pd reactions results in a decrease in the resonance frequency, F, of the coated oscillator. The experiments were performed using a frequency shift detection technique that, according to several authors (e.g., D'Amico et al., 1982, 1986), is the most suitable for practical applications because of its sensitivity and simplicity. The Pd–SAW hydrogen detector fabricated by D'Amico et al. (1986) has the same characteristics as the one presented earlier by the same authors (D'Amico et al., 1982/83) (see Table 7.1) with the exception of different Pd thicknesses ($L = 1900$, 3800, and 7600 Å). Figure 7.13 shows the variation of $\Delta F/F$ as a function of Pd thickness, L, for two different H$_2$ concentrations; $\Delta F/F$ increases with increasing L. On the other hand, it can be seen that the relative frequency increase with H$_2$ concentration is not drastically influenced by the concentration, for a given Pd thickness.

Figure 7.14 shows the variation of the response time, R_s, as a function of the flow rate for three different H$_2$ concentrations. The flow rate does influence the response time of the sensor, which increases monotonically with decreasing flow rate between 810 and 8100 mL/min. D'Amico et al. (1982) have also shown that there is an optimum Pd thickness for which R_s is minimum. According to these authors, among three thicknesses of Pd coating, 1900, 3800, and 7600 Å, it is the intermediate one (3800 Å) that presented the best Pd–SAW response, while the thinnest one presented the worst response.

The main results of the SAW hydrogen detection can be summarized as follows:

a. Sensitivity Limit. As is shown in Eqs. (7.4)–(7.6) the sensitivity and the size of the SAW detectors are directly related to their operating resonance

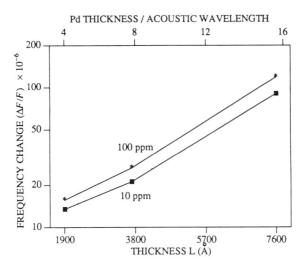

Figure 7.13. Relative frequency change vs. Pd thickness for two different concentrations of hydrogen in nitrogen for the Pd–SAW sensor fabricated by D'Amico et al. (1982).

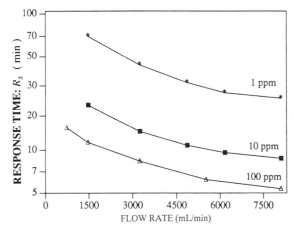

Figure 7.14. Variation of the SAW response time, R_s, as a function of flow rate for three different hydrogen concentrations as measured for Pd film thickness $L = 7600$ Å; $T = 20\,°C$ (D'Amico et al., 1982).

frequency. Thus, a device operating at 300 MHz presents a frequency shift 100 times greater than a 30 MHz device for the same absorbed gas mass [see Eq. (7.8)]. However, the baseline noise for the 300 MHz SAW device is higher than that for the device operating at 30 MHz (Wohltjen, 1984). The reduction of the device area has implications with respect to the minimum mass change

Table 7.5. Estimated SAW Device Sensor Performance

Frequency (MHz)	Device Area (cm^2)	Baseline Noise (Hz)	Minimum Detectable Mass Change (g)
30	1	3	3×10^{-9}
300	10^{-2}	30	3×10^{-12}
3000	10^{-4}	300	3×10^{-15}

Source: Wohltjen (1984).

that can be detected by the SAW delay-line oscillator. Some estimates of the SAW performances at various operating conditions are presented in Table 7.5.

b. Influence of the Device Substrate. Figure 7.15 shows $\Delta F/F$ values as obtained upon exposure to 1% of hydrogen in nitrogen as a function of the thickness of the palladium film normalized to the acoustic wavelength. It can be seen that the Pd/ZnO/SiO$_2$/Si structure shows higher response than the Pd/STX-SiO$_2$ and Pd/YZ-LiNbO$_3$ structures.

c. Temperature Effect on SAW Detectors. An undesirable property of

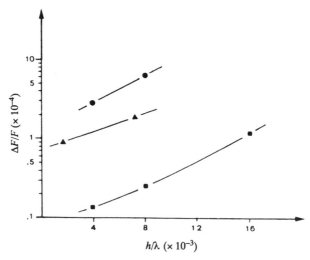

Figure 7.15. Fractional frequency change as a function of the normalized palladium film thickness upon exposure to a concentration of 1% of hydrogen in nitrogen, as measured on three different substrate structures: (●) Pd/ZnO/SiO$_2$/Si; (▲) Pd/STX-SiO$_2$; and (■) Pd/YZ-LiNbO$_3$ (Caliendo et al., 1988).

$LiNbO_3$ SAW oscillators is their instability at high temperatures. According to Joshi and Brace (1985), the resonant frequency is found to decrease linearly with temperature with a slope of $-6800\,Hz/°C$ at $F = 75\,MHz$, whereas Ricco et al. (1985) found this frequency shift to be $-8800\,Hz/°C$ at a resonance frequency of 110 MHz.

   d. Comparison Between SAW and PQCMB Detectors. It is well known that SAW and piezoelectric quartz crystal microbalance (PQCMB) detectors are mass sensitive and require a selective coating layer. These two detectors use a shift in acoustic wave resonant frequency (ΔF) as a signal. However, these two devices have two fundamental differences: (a) the resonant frequency of the SAW device can be a few hundred times higher than that of the PQCMB detector, which implies greater sensitivity; (b) the SAW device has the ability to occupy much less space than the PQCMB device. The reduction of device area has implications with regard to the minimum mass change that can be detected by a SAW delay line oscillator. According to Wohltjen (1984), for a YX-quartz SAW delay-line oscillator Eq. (7.8) can be written as

$$\Delta F = 1.3 \times 10^6 F^2 (\Delta M / A_s) \tag{7.11}$$

where ΔM is an grams and A_s is in square centimeters. It should be noted that the numerical factor of Eq. (7.11) is smaller than that of Eq. (6.47) for the PQCMB detector. However, since the operating frequency of the SAW device is much greater than that of the bulk wave detector, a compensation in sensitivity occurs. A direct comparison between a 6 MHz bulk wave oscillator and a 75 MHz SAW oscillator shows that the SAW device theoretically should produce a much higher frequency shift. Actually, by term-to-term division of Eqs. (7.11) and (6.47) we have (Christofides and Mandelis, 1990)

$$\frac{\Delta F_{SAW}}{\Delta F_{PQCMB}} = \frac{1.3 \times 10^6}{2.3 \times 10^6} \frac{75^2}{6^2} \approx 88.3 \tag{7.12}$$

Equation (7.12) shows that the sensitivity of the SAW device is 88 times greater than that of the PQCMB device. This theoretical value was experimentally confirmed by the comparison between results of D'Amico et al. (1982/83) for the Pd–SAW detector and those of Christofides and Mandelis (1989) for the Pd–PQCMB device. Sensitivity problems of the SAW devices have also been studied by Bryant et al. (1983).

7.4.2. Detection of Ammonia

D'Amico et al. (1987) used a SAW device (STX–SiO_2, 30 MHz) coated with platinum for the detection of ammonia. The response of the SAW device as

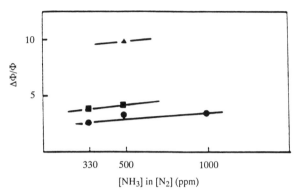

Figure 7.16. Response of the SAW sensor as a function of the ammonia concentration for various Pt coating thicknesses: (●) 900 Å; (■) 1300 Å; and (▲) 2700 Å (D'Amico et al., 1987).

a function of NH_3 concentration has been investigated for various thicknesses of the chemical layer.

The measurements have been performed at 25 °C in dynamic flow by using pure nitrogen as a balance gas. The response and recovery times have been found to be on the order of a few minutes. Figure 7.16 presents the fractional change of the phase delay as a function of $[NH_3]$ for three different platinum thicknesses. As in the case of the SAW hydrogen detection (see Fig. 7.13), the sensitivity increases with the thickness of the chemical layer. D'Amico and co-workers (1987) claim a linear response of their device in the presence of the ammonia gas. Nevertheless, the small amount of data, as well as the limited range of concentrations studied, do not allow any secure extrapolation of the foregoing conclusion to a larger concentration range.

D'Amico et al. (1987) have also performed some measurements at 35 °C that did not show any significant change in the sensitivity of the device. In conclusion, it seems that the development of SAW toward the monitoring of ammonia concentrations is very possible and the first results obtained by the aforementioned authors are promising. The same group (Arya et al., 1988) have also obtained some preliminary ammonia measurements by using a ZnO-based device.

7.4.3. Detection of Carbon Dioxide

The application of SAW to the detection of carbon dioxide (Table 7.3) is of great interest in the field of SAW gas sensors. Various chemical coatings have been used in order to reach high sensitivity, selectivity, and reversibility. Some positive results have been obtained, but it seems that researchers are still trying to improve and optimize various types of chemical layers. Nieuwenhuizen and Nederlof (1990) have used several amines and imines as well as zeolites with the goal of obtaining selective interaction with CO_2. Use of zeolites has proved to be very problematic because the adhesion of these compounds to the substrate is not easy. These authors further used poly(ethyleneine) (PEI) as a 50% solution in water. The SAW frequency of their device was 40 MHz. Some experiments at various temperatures show that the detection of CO_2 is strongly dependent on temperature. Moreover, it has been found that some hysteresis phenomena with respect to temperature are also present. Figure 7.17 presents the SAW response to carbon dioxide concentration at 70 °C. By reviewing the small amount of work done on SAW detection of CO_2 (see Table 7.3), we can make the following comments:

a. The oxygen in air is partially responsible for a temporal deterioration of the device's CO_2-sensing abilities owing to some chemical interactions between O_2 and PEI.

b. Because of the acid–base interactions of CO_2 with PEI, the SAW response is highly affected in the presence of H_2O traces.

c. Interferences from H_2O, O_2, NH_3, and ethanol have been observed.

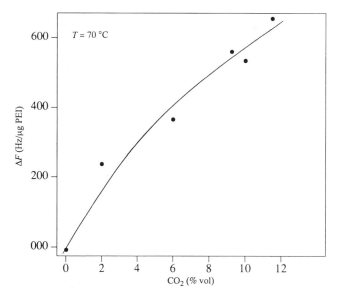

Figure 7.17. The response of SAW sensor with freshly prepared PEI layers to varying CO_2 concentrations at 70 °C (Nieuwenhuizen and Nederlof, 1990).

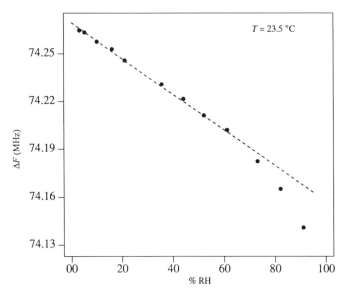

Figure 7.18. Variation of SAW oscillator frequency as a function of relative humidity for a polyimide-coated device at 23.5 °C (Joshi and Brace, 1985).

7.4.4. Detection of Humidity

Joshi and Brace (1985) used a 75 MHz SAW oscillator fabricated on YZ-LiNbO$_3$ substrate and performed simultaneous measurements of relative humidity (RH) and temperature. The SAW device was coated with a hygroscopic polyimide polymer film that has proven to be selective, very sensitive, and with fast response. As shown in Fig. 7.18, the frequency change ΔF varies by more than 150 kHz as RH varies from 0% to 100%. We note that the SAW response is linear between 0% and 60% but follows a nonlinear shape at higher RH. A large hysteresis has also been found in the humidity response of the polyimide-coated surface acoustic device. Huang (1987) used a SAW device to study the detection of RH under various frequencies.

7.4.5. Detection of Other Gases

In this subsection we shall highlight several findings (of authors cited in Table 7.3) that have been discovered during the surface acoustic wave detection process of various gases such as NH$_3$, styrene vapors, derivatives of nitrobenzene, NO$_2$, and toluene:

a. Zellers et al. (1990) used a SAW sensor with an organoplatinum coating for selective real-time measurements of styrene vapors. The device exhibited reversible response, with butadiene being the only exception. It was also found that atmospheric humidity influences the SAW response, a fact consistent with Joshi and Brace's (1985) work.

b. A number of derivatives of nitrobenzene have been detected with a SAW device by Heckl et al. (1990). The sensor was coated with aminopropyltriethoxysilane (APTES). Figure 7.19 presents the frequency shift vs. time after multiple injections of the analytes on the APTES-coated SAW layers.

c. A special effort has been made by several workers toward the detection of NO_2 (see Table 7.3). In 1989 Nieuwenhuizen et al. used two identical $ZnO-SiO_2-Si$ layered delay lines for the detection of nitrogen dioxide. Table 7.6 shows the sensitivity and selectivity of several types of SAW detectors for the detection of NO_2, NH_3, and H_2O at 150 °C.

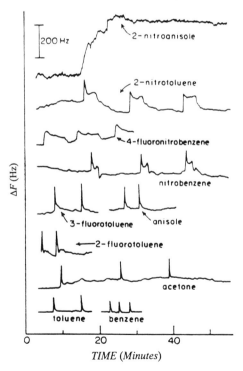

Figure 7.19. Frequency shift as a function of time after multiple injections of the analytes on the APTES-coated SAW surface (Heckl et al., 1990).

Table 7.6. Sensitivity (S) of Various Types of Sensors to NO_2, NH_3, and H_2O at 150 °C

Code	Frequency (MHz)	NO_2	NH_3	H_2O
CuPC134[a]	70	200.0	−1.0	0.0
CuPC135[a]	70	920.0	−5.2	0.0
ZnO[a]	70	40.0	−0.7	−0.1
CuPC65	52	19.8	−1.6	0.0
CuPC76	80	−74.0	3.7	0.0
CuPC187	52	0.3	−0.6	0.003
Quartz[a]	80	0.0	0.0	0.0

Source: Nieuwenhuizen et al. (1989).
[a]Single delay-line measurements.

d. In general, as in the case of the bulk piezoelectric device, a lack of absolute selectivity during the SAW detection of gases causes several interferences.

e. Wenzel and White (1989a) described the use of flexural plate waves traveling in thin composite plates of $ZnO/Al/Si_xN_y$ for gravimetric

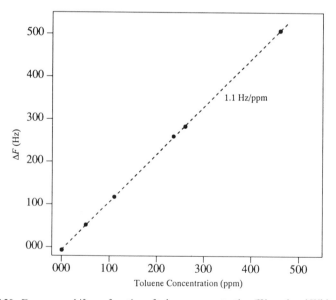

Figure 7.20. Frequency shift as a function of toluene concentration (Wenzel and White, 1989b).

chemical vapor sensors. They used poly(dimethylsiloxane) and ethyl cellulose as chemical coatings. These authors detected toluene, 1,1,1,-trichloroethane, and CCl_4 in ppb ranges. Figure 7.20 shows the linearity of the response to toluene vapor.

7.5. ANALYTICAL COMPARISON OF THE MAIN PIEZOELECTRIC DEVICES

Wenzel and White (1989a, b) gave a critical comparison of the sensitivity of the PQCMB, SAW, and flexural plate-mode-wave (PMW) ultrasonic gravimetric sensors. These authors proved that for the three aforementioned devices the mass sensitivities are as follows:

$$S_{PQCMB} = -\frac{1}{\rho\lambda} \tag{7.13}$$

$$S_{SAW} = -\frac{K_\sigma}{\rho\lambda} \tag{7.14}$$

and

$$S_{PMW} = -\frac{1}{2\rho\lambda} \tag{7.15}$$

where ρ is the density; λ is the half-wavelength; K_σ ranges from 1 to 2; and d is the material thickness. In the case of the plate-mode sensor $d \ll \lambda$. From Eq. (7.13) it is obvious that, in order to increase the sensitivity of the PQCMB or SAW, the wavelength must be decreased, which means an increase of operating frequency since $F = V/\lambda$, where V is the phase velocity. On the other hand, the plate-mode sensitivity, S_{PMW}, can be increased independently of the wavelength just by reducing the plate thickness. Finally, by summarizing the results of Wenzel and White (1989a, b), we can conclude that the structure of the plate-mode sensor permits the best sensitivity among the three devices at lower operating frequencies.

7.6. ARRAY OF SAW DEVICES: PATTERN RECOGNITION

As in the case of the PQCMB, the need for devices with high selectivity has led some workers in the field of SAW devices to build arrays by using several of those devices coated with various coatings.

Ballantine et al. (1986, p. 3059) used 12 SAW devices coated with different chemical layers, which were then exposed to 11 vapors. The data were analyzed with a certain statistical pattern recognition technique described as follows:

> The sensors encode chemical information about the vapor in numerical form. Each sensor defines an axis in a multidimensional space. Vapors can be represented as points positioned in this space according to sensor responses. Vapors that produce similar responses from the set of coatings will tend to cluster near one another in space. Pattern recognition uses multivariate statistics and numerical analysis to investigate such clustering and to elucidate relationships in multidimensional data sets without human bias. In addition this method can reduce interference effects and improve selectivity in analytical measurements.

7.7. OTHER APPLICATIONS OF SAW DEVICES

Gas detection is not the only application of SAW devices. In several other fields SAW devices have also been used, such as in analytical chemistry, in gas chromatography, and as mechanical sensors. Table 7.7 lists various applications of the piezoelectric crystal and cites the authors who developed these techniques. We close this chapter with a brief overview of some noteworthy nongaseous applications of SAW devices.

The dependence of wave velocities on temperature led to a temperature sensor based on a SAW delay-line oscillator with good linearity (Hauden et al., 1981b). The thermal properties of quartz substrate were investigated in order to find a propagation geometry that allows large and linear temperature coefficient of the SAW delay. It is also well known that SAW velocities can be strongly affected by stressing the crystal through which the wave travels (Hauden et al., 1981a). This property led to the first SAW pressure sensor, developed by Cullen and Reeder (1975). Toda and Mizutani (1983) and Joshi (1983) reported the first SAW electric field sensor. They exploited the fact that an electrical field normal to the piezoelectric surface along which a SAW is propagating causes a shift in the wave velocity because of the change in the stiffness of the crystal. Requirements for a high-quality SAW electric sensor are that the relative change in phase velocity vs. electric field must be as high and linear as possible. SAW electric sensors are suitable for dc measurements. The measurements of ac voltage are also possible provided, of course, that the delay of the acoustic wave is smaller than the ac period. Inaba et al. (1982) reported the first SAW voltmeter. Hanna (1987) reported the SAW signal dependence on an applied magnetic field.

The SAW properties have yielded many other kinds of sensors in recent years, such as SAW acceleration, displacement, and flow sensors, as well as

Table 7.7. Various Applications of SAW Devices

References	SAW Sensors
Hauden et al. (1981a, b)	Temperature
Bao et al. (1987)	Temperature
Cullen and Reeder (1975)	Pressure
Cullen and Montress (1980)	Pressure
Staples et al. (1981)	Pressure
Risch (1984)	Pressure
Toda and Mizutani (1983)	Electric field
Joshi (1983)	Electric field
Inaba et al. (1982)	Voltage
Joshi (1982)	Voltage
Gatti et al. (1983a, b)	Voltage
Palma et al. (1984)	Voltage
Palmieri et al. (1986a)	Voltage
Palmieri et al. (1986b)	Voltage
Hanna (1987)	Magnetic field
Tiersten et al. (1980)	Accelaration
Staples et al. (1981)	Accelaration
Hartemann and Meunier (1981)	Accelaration
Meunier and Hartemann (1982)	Accelaration
Hartemann and Meunier (1983)	Accelaration
Hauden et al. (1985)	Accelaration
Hauden and Loewenguth (1985)	Accelaration
Ishido et al. (1987)	Displacement
Ahmad (1985)	Flow Meter
Lao (1980)	Gyro effect (rotation)
Tiersten et al. (1980)	Gyro effect (rotation)
Kornovich and Harnak (1977)	Film-thickness measurements
Adler et al. (1980)	Acoustoelectric effect
Wohltjen and Dessy (1979)	Chemical Analysis
Wohltjen and Dessy (1979)	Gas chromatography
Brace et al. (1987)	Polymer–metal studies
Dieulesant et al. (1987)	Liquid-level meter
Martin et al. (1987a, b)	Sensing in liquids

a rotation (gyro-effect) SAW detector. In 1977 Kornovich and Harnak used a SAW device to measure the thickness of metal films deposited in a vacuum evaporation apparatus. Adler et al. (1980) have also used a surface acoustic device to measure the mobility of charge carriers in a semiconducting film by means of the acoustoelectric effect. Chemical studies of polymeric films have also been performed by using a SAW device (Brace et al., 1987). These authors studied polymer–water interactions using SAW and investigated the thermodynamics and kinetics of the system. Dieulesant et al. (1987) used the SAW structure for the development of a liquid level sensor. Martin et al. (1987a, b) developed a SAW device that allows highly accurate measurements of mass changes on a surface immersed in a liquid. From the foregoing, it is apparent that many new and exciting applications of SAW technologies in the fields of gaseous and condensed matter sensing, lie ahead in the years to come.

REFERENCES

Adler, R., Janes, D., Datt, S., and Hunsinger, B.J. (1980). *Proc. IEEE Ultrason. Symp.*, Boston, *1980*, p. 139.

Ahmad, N. (1985). *Proc. IEEE Ultrason. Symp.*, San Francisco, *1985*, Vol. 1, p. 483.

Anisimkin, V.I., Verona, E., and D'Amico, A. (1989). *Nuovo Cimento, Note Brevi* **11D**, 503.

Arya, S.P.S., D'Amico, A., and Verona E. (1988). *Thin Solid Films* **157**, 169.

Auld, B.A. (1973). *Acoustic Fields and Waves in Solids*, Vol. 2, Chapter 12. Wiley (Interscience), New York.

Ballantine, D.S., Rose, S.L., Grate, J.W., and Wohltjen, H. (1986). *Anal. Chem.* **58**, 3058.

Bao, X.Q., Burkhard, W., Varadan, V.V., and Varadan, V.K. (1987). *Proc. IEEE Ultrason. Symp.*, Denver, *1987*, Vol. 1, p. 583.

Barensz, A.W., Vis, J.C., Nieuwenhuizen, M.S., Nieuwkoop, E., Vellekoop, M.J., Ghijsen, W.J., and Venema, A. (1985). *Proc. IEEE Ultrason. Symp.*, San Francisco, 1985, Vol. 1, p. 586.

Brace, J.G., Sanfelippo, T.S., and Joshi, S.G. (1987). *Proc. Int. Conf. Solid-State Sens. Actuators, Transducers '87, 4th*, Tokyo, p. 467.

Brace, J.G., Sanfelippo, T.S., and Joshi, S.G. (1988). *Sens. Actuators* **14**, 47.

Bryant, A., Lee, D.L., and Vitelino, J.F. (1981). *Proc. IEEE Ultrason. Symp.*, Chicago, *1981*, p. 171.

Bryant, A., Poirier, M., Riley, G., Lee, D.L., and Vitelino, J.F. (1983). *Sens. Actuators* **4**, 105.

Caliendo, C., D'Amico, A., Verardi, P., and Verona, E. (1988). *Proc. IEEE Ultrason. Symp.*, Chicago, *1988*, p. 569.

Campbell, C. (1989). *Surface Acoustic Wave Devices and Their Signal Processing Applications.* Academic Press, Boston.

Christofides, C., and Mandelis, A. (1989). *J. Appl. Phys.* **66**, 3986.

Christofides, C., and Mandelis, A. (1990). *J. Appl. Phys.* **68**, R1.

Chuang, C.T., and White, R.M. (1981). *Proc. IEEE Ultrason. Symp.*, Chicago, *1981*, p. 159.

Cullen, D.E., and Montress, G.K. (1980). *Proc. IEEE Ultrason. Symp.*, Boston, *1980*, p. 696.

Cullen, D.E., and Reeder, T.M. (1975). *Proc. IEEE Ultrason. Symp.*, Los Angeles, *1975*, p. 519.

D'Amico, A., and Verona, E. (1988). *Prog. Solid State Chem.* **18**, 177.

D'Amico, A., and Verona, E. (1989). *Sens. Actuators* **17**, 55.

D'Amico, A., Palma, A., and Verona, E. (1982). *Appl. Phys. Lett.* **41**, 300.

D'Amico, A., Palma, A., and Verona, E. (1982/83). *Sens. Actuators* **3**, 31.

D'Amico, A., Gentili, M., Verardi, P., and Verona, E. (1986). *Proc. Int. Meet. Chem. Sens. 2nd*, Bordeaux, Fr., p. 743.

D'Amico, A., Petri, A., Verardi, P., and Verona, E. (1987). *Proc. IEEE Ultrason. Symp.*, Denver, *1987*, Vol. 1, p. 633.

Dieulesaint, E., Royer, D., Legras, O., and Boubeniden, F. (1987). *Proc. IEEE Ultrason. Symp.*, Denver, *1987*, Vol. 1, p. 569.

Gatti, E., Palma, A., and Verona, E. (1983a). *Sens. Actuators* **4**, 45.

Gatti, E., Palma, A., and Verona, E. (1983b). *Proc. IEEE Ultrason. Symp.*, Atlanta, *1983*, p. 291.

Hanna, S.W. (1987). *IEEE Trans. Ultrason., Ferroelectr., Freq. Control* **UFFC-2**, 191.

Hartemann, P., and Meunier, P.L. (1981). *Proc. IEEE Ultrason. Symp.*, Chicago, *1981*, Vol. 1, p. 152.

Hartemann, P., and Meunier, P.L. (1983). *Proc. IEEE Ultrason. Symp.*, Atlanta, *1983*, p. 291.

Hauden, D., and Loewenguth, B. (1985). *Appl. Phys. Lett.* **47**, 1271.

Hauden, D., Jaillet, G., and Coquerel, R. (1981a). *Proc. IEEE Ultrason. Symp.*, Chicago, *1981*, p. 148.

Hauden, D., Planat, M., and Gagnepain, J.J. (1981b). *IEEE Trans. Sonics Ultrason.* **SU-28**, 342.

Hauden, D., Bindler, F., and Coquerel, R. (1985). *Proc. IEEE Ultrason. Symp.*, San Francisco, *1985*, Vol. 1, p. 486.

Heckl, W.M., Marassi, F.M., Kallury, K.M.R., Stone, D.C., and Thompson, M. (1990). *Anal. Chem.* **62**, 32.

Huang, P.H. (1987). *Proc. Int. Conf. Solid-State Sens. Actuators, Transduers '87, 4th*, Tokyo, p. 462.

Inaba, R., Kasahara, Y., and Wasa, K. (1982). *Proc. IEEE Ultrason. Symp.*, San Diego, *1982*, p. 312.

Ishido, M., Imaizumi, T., and Toyoda, M. (1987). *IEEE Trans. Instrum. Meas.* **IM-36**, 83.

Joshi, S.G. (1982). *Proc. IEEE Ultrason. Symp.*, San Diego, *1982*, p. 317.

Joshi, S.G. (1983). *Rev. Sci. Instrum.* **54**, 1012.

Joshi, S.G., and Brace, J.G. (1985). *Proc. IEEE Ultrason. Symp.*, San Francisco, *1985*, Vol. 1, p. 600.

Kornovich, S., and Harnak, E. (1977). *Rev. Sci. Instrum.* **48**, 920.

Lao, B.Y. (1980). *Proc. IEEE Ultrason. Symp.*, Boston, *1980*, p. 687.

Lord Rayleigh (1885). *London Math. Soc. Proc.* **17**, 4.

Martin, S.J., Ricco, A.J., Ginley, D.S., and Zipperian, T.E. (1987a). *IEEE Trans. Ultrason., Ferroelectr., Freq. Control* **UFFC-34**, 143.

Martin, S.J., Ricco, A.J., and Hughes, R.C. (1987b). *Proc. Int. Conf. Solid-State Sens. Actuators, Transducers '87, 4th*, Tokyo, p. 478.

Matthews, H. (1977). *Surface Wave Filters.* Wiley, New York.

Meunier, P.L., and Hartemann, P. (1982). *Proc. IEEE Ultrason. Symp.*, San Diego, *1982*, p. 299.

Morgan, D.P. (1973). *Ultrasonics* **11**, 121.

Nieuwenhuizen, M.S., Nederlof, A.J., Vellekoop, M.J., and Venema, A. (1989). *Sens. Actuators* **19**, 385.

Nieuwenhuizen, M.S., and Nederlof, A.J. (1990). *Sens. Actuators B* **2**, 97.

Oliner, A.A. (1978). *Surface Acoustic Waves.* Springer-Verlag, New York.

Palma, A., Palmieri, L., Socino, G., and Verona, E. (1984). *Proc IEEE Ultrason. Symp.*, Dallas, *1984*, Vol. 2, p. 951.

Palmieri, L., Socino, G., and Verona, E. (1986a). *Appl. Phys. Lett.* **23**, 1581.

Palmieri, L., Socino, G., and Verona, E. (1986b). *Proc. IEEE Ultrason. Symp.*, Williamsburg, VA, *1986*, Vol. 2, p. 1093.

Ricco, A.J., and Martin, A.J. (1988). *Proc. 172nd Electrochem. Soc. Meet.* **87-13**.

Ricco, A.J., Martin, S.J., and Zipperian, T.E. (1985) *Sens. Actuators* **8**, 319.

Ricco, A.J., Martin, S.J., Frye, G.C., and Niemczyk, T.M. (1988). *IEEE Workshop Solid State Sens. Actuators*, Head Island, SC, p. 23.

Risch, M.R. (1984). *Sens. Actuators* **6**, 127.

Proceedings of the IEEE (1976). *Special Issue on Surface Acoustic Wave Devices and Applications*, Vol. 64. Inst. Electr. Electron. Eng., New York.

Special Issue on Microwave Acoustics (1969). *IEEE Trans. Microwave Theory Tech.* **MTT-17**.

Staples, E.J., Wise, J., and DeWames, R.E. (1981). *Proc. IEEE Ultrason. Symp.*, Chicago, *1981*, p. 155.

Tiersten, H.F., Stevens, D.S., and Das, P.K. (1980). *Proc. IEEE Ultrason. Symp.*, Boston, *1980*, p. 692.

Thompson, M., Dhaliwal, G.K., Arthur, C.L., and Calabrese, G.L. (1987). *IEEE Trans. Ultrason., Ferroelectr., Freq. Control* **UFFC-34**, 127.

Toda, K., and Mizutani, K. (1983). *J. Acoust. Soc. Am.* **74**, 667.

Vellekoop, M.J., Nieuwkoop, E., Haartsen, J.C., and Venema, A. (1987). *Proc. IEEE Ultrason. Symp.*, Denver, *1987*, Vol. 1, p. 64.

Venema, A., Nieuwkoop, E., Vellekoop, M.J., Nieuwenhuizen, M.S., and Barendsz, A.W. (1986). *Sens. Actuators* **10**, 47.

Venema, A., Nieuwkoop, E., Vellekoop, M.J., Ghijsen, W.J. Barendsz, A.W., and Nieuwenhuizen, M.S. (1987). *IEEE Trans. Ultrason., Ferroelectr., Freq. Control* **UFFC-34**, 148.

Verona, E. (1988). In *Third Course on Ultrasonic Signal Processing* (A. Alippi and G. Scarano, Eds.), p. 199. International School of Physical Acoustics, World Scientific, Erice, Italy.

Vetelino, J.F., Lad, R., and Falconer, R.S. (1986). *Proc. IEEE Ultrason. Symp.*, Annapolis, MD, p. 549.

Vetelino, J.F., Lad, R.K., and Falconer, R.S. (1987). *Trans. Ultrason., Ferroelectr., Freq. Control* **UFFC-34**, 156.

Wenzel, S.W., and White, R.M. (1988a). *IEEE Workshop Solid State Sens. Actuators*, Hilton Head Island, SC, p. 27.

Wenzel, S.W., and White, R.M. (1988b). *IEEE Trans. Electron Devices* **ED-35**, 735.

Wenzel, S.W., and White, R.M. (1989a). *Proc. IEEE Ultrason. Symp.*, Montreal, *1989*, Vol. 1, p. 595.

Wenzel, S.W., and White, R.M. (1989b). *Appl. Phys. Lett.* **54**, 1976.

White, R.M., and Voltmer, F.W. (1965). *Appl. Phys. Lett.* **7**, 314.

White, R.M., Wicher, P.J., Wenzel, S.W., and Zellers, E.T. (1987). *IEEE Trans. Ultrason., Ferroelectr., Freq. Control* **UFFC-34**, 162.

Wohltjen, H. (1984). *Sens. Actuators* **5**, 307.

Wohltjen, H., and Dessy, R.E. (1979). *Anal. Chem.* **51**, 1458.

Wohltjen, H., Snow, A., and Ballantine, D. (1985). *Proc. IEEE Ultrason. Symp.*, San Francisco, *1985*, p. 66.

Wohltjen, H.W., Snow, A.W., Barger, W.R., and Ballantine, D.S. (1987). *IEEE Trans. Ultrason., Ferroelectr., Freq. Control* **UFFC-34**, 172.

Zellers, E.T., White, R.M., Rappaport, S.M., and Wenzel, S.W. (1987). *Proc. IEEE Ultrason. Symp.*, Denver, *1987*, Vol. 1, p. 459.

Zellers, E.T., Hassold, N., White, R.M., and Rappaport, S.M. (1990). *Anal. Chem.* **62**, 1227.

Zemel, J.N. (1990). *Rev. Sci. Instrum.* **61**, 1579.

CHAPTER

8

PYROELECTRIC AND THERMAL SENSORS

8.1. INTRODUCTION AND HISTORICAL PERSPECTIVE

8.1.1. Pyroelectricity: A 2300-Year History

The phenomenon of pyroelectricity is not new; it has some 2300 years of history. The first description of the pyroelectric effect was published by the Greek philosopher Theophrastus (ca. 372–287 B.C.) in his book *On Stones* (Lang, 1974). Nevertheless, the phenomenon of pyroelectricity was not only a hot subject in the literature of the Greek and Roman periods but also during the Dark and Middle Ages. It is interesting to note that a book printed in 1497 entitled *Hortus Sanitatis Major* contains several chapters describing minerals that very likely possess pyroelectric properties (Lang, 1974). Two thousand years after the "father" of pyroelectricity, Theophrastus, the phenomenon was "rediscovered" by Dutch gem cutters in 1703, and the property of pyroelectricity as found in a stone brought by Dutch sailors from the East Indies was first described in a scientific journal of the time, *Histoire de l'Académie Royale des Sciences*, by Louis Lemery in 1717. The first serious scientific investigation was presented to the Royal Academy of Sciences in Berlin by Aepinus in 1756. In the nineteenth century David Brewster was the first author to use the term *pyroelectricity* in a paper published in 1824 entitled, "Observations of the Pyro-Electricity of Minerals." It is unknown just how or why Brewster came to use the foregoing term for this phenomenon. In fact, etymologically the name is derived from the Greek work $\pi v \rho$ (*pir*), which means fire. It is likely that an association between "fire" and "heat" was made. Jean-Mothée Gaugain in 1856 made the first precise measurements of pyroelectric charges with a self-discharging electroscope, and the first major scientific analysis of the pyroelectric phenomenon was described in 1878 and 1893 by William Thomson (Lord Kelvin), who associated the electric field with the pyroelectric effect. Later, Charles Friedel and Jacques Curie recognized the difference between uniform and nonuniform heating. As was cited in Chapter 6, this observation led Jacques and Pierre Curie to their famous discovery of piezoelectricity. Measurements of the pyroelectric effect first appeared shortly before World War I (Roentgen, 1914). Important theoretical work was also carried out in the early twentieth century. In 1921/22 Max Born published some papers concerning the interpretation of pyroelectricity by means of lattice dynamics. In 1938 the key paper that led pyroelectricity to its recent developments was published by Yeou Ta (1938) and was entitled, "Action of Radiation on Pyroelectric Crystal." This paper was the basis of the development of photopyroelectric infrared detectors as we know them today. Several theoretical and experimental studies of the pyroelectric effect have been performed during the last 20 years. The first patent for a pyroelectric device was obtained by Leon J. Sivian in 1950, and Rossetti constructed a detector using lithium sulfate and potassium tartate crystals 6 years later.

8.1.2. Pyroelectricity: Introduction to Gas Sensors

Pyroelectricity is the manifestation of the spontaneous polarization dependence on temperature of certain anisotropic solids. It is important to note that all the ferroelectric crystals possess piezoelectric and pyroelectric properties. All pyroelectric crystals are also piezoelectric, albeit not all piezoelectric crystals are pyroelectric; for example, quartz and GaAs are both piezoelectric, but neither is pyroelectric. This effect is exhibited only by solids (single crystal, ceramic, or polymer) that satisfy crystallographic requirements such as (a) the crystal lattice must have no center of symmetry and (b) the crystal must have no more than one axis of rotational symmetry. Microscopically, the pyroelectric (PE) effect takes place because of the asymmetric environment experienced by electrically charged species within the crystal structure of the material. A pyroelectric material becomes electrically neutral, when in a constant temperature environment for a period of time. If there is a small change of the temperature, ΔT, the pyroelectric material becomes electrically polarized and a voltage arises between certain directions in the material (Putley, 1970). While thermal sensitivity is very important, these materials must have a very weak dependence of their pyroelectric property on temperature over a wide temperature range that includes the working range. As Zemel (1990) has pointed out, in order to ensure this the PE material must have sufficiently high Curie temperature, T_C. As is shown in Fig. 8.1, the PE coefficient $p(T)$ vs. temperature is fairly independent of temperature in the range 0–300 °C.

Zemel et al. (1981) and Young (1982) reported the first pyroelectric chemical sensors ($Pd–LiTaO_3$). Except for a few more studies reported by Hall et al. (1984); D'Amico et al. (1985a, b); Zemel (1985); D'Amico and Zemel (1985); and Aihara and Kawakami (1985), the pyroelectric phenomenon did not appear to generate much interest in the chemical detector community; thus the pyroelectric chemical sensor based on $Pd–LiTaO_3$ was not further

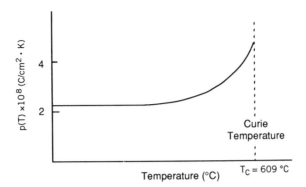

Figure 8.1. Representative temperature dependence of the PE coefficient of $LiTaO_3$; T_C is the Curie temperature (608 °C), where the material undergoes a first-order phase transition (Zemel, 1990).

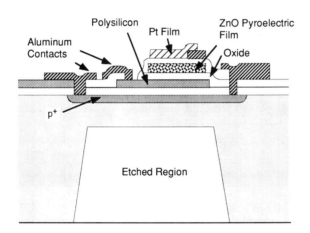

Figure 8.2. ZnO pyroelectric sensor component of an integrated thermal PE sensor. The p^+ region was used as a heater to calibrate the device and drive-in of absorbed gases (Polla et al., 1985).

developed. Polla et al. (1985) employed a thin-film ZnO pyroelectric to study its sensitivity to gases. Figure 8.2 shows the basic configuration of the ZnO–PE sensors. Recently an experimental effort by the present authors has led to the development of a promising new simple, sensitive, and fast photo-pyroelectric (PPE) chemical detector based on poly(vinylidene fluoride) (PVDF): the ac Pd–PVDF photopyroelectric sensor for H_2 gas detection

(Mandelis and Christofides, 1989, 1990a, 1991; Christofides and Mandelis, 1989a, b).

This chapter describes the operating characteristics of pyroelectric (and thermal) gas sensors. All these detectors were reported coated with palladium or platinum electrodes, which are well known for their ability to selectively adsorb hydrogen. Exposure to hydrogen gas was shown to produce a signal difference between the Pd and reference electrodes. In fact, in this chapter two main types of pyroelectric detectors are discussed (a) dc pyroelectric chip detectors; and (b) ac photopyroelectric thin-film sensors. The fundamental difference between the operating mechanisms of the dc and ac pyroelectric sensors is that for the dc detector the shift on the Pd–LiTaO$_3$ pyroelectric signal is due to the change of temperature resulting from the deposition of adsorption energy (Zemel, 1985), whereas for the ac sensor the response is due to different mechanisms (Mandelis and Christofides, 1991). A comparison of ac and dc pyroelectric devices will be presented with respect to their sensitivity, chemical selectivity, and stability. Finally, the Pd–mica–Au capacitive thermal wave sensor will be described.

8.2. THEORY OF PYROELECTRIC SIGNAL GENERATION

In pyroelectric materials the polarization vector is a function of temperature. As the temperature varies, the surface charge also varies owing to dimensional changes in the pyroelectric. This property results in a potential difference between the two opposite surfaces of the material, generally known as pyroelectricity. According to Zemel (1985), the charge density, Q, on a pyro-electric capacitor may be written in the following form:

$$Q = -C_p(V - V_0) + p(T)(\langle T \rangle - T_0) \tag{8.1}$$

where $V - V_0$ is the potential drop across the pyroelectric capasitor; C_p, the capacitance; p, the pyroelectric coefficient; T, the instantaneous temperature; T_0, the initial temperature; and $\langle T \rangle$, the instantaneous temperature rise of the pyroelectric averaged over its thickness. Therefore, the total pyroelectric current will be the integral of the current density over the capacitor surface, A. Figure 8.3 shows an equivalent circuit where the pyroelectric element is described as ideal source (Coufal et al., 1987). Using Eq. (8.1) one can write

$$I = \int_A \frac{dQ}{dt} dA = \int_A \left(-C_p A \frac{dV}{dt} + p \frac{d\langle T \rangle}{dt} \right) dA \tag{8.2}$$

The output voltage measured under load conditions may be written as (Zemel,

(a)   (b)

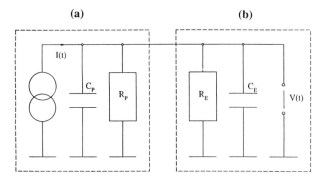

Figure 8.3. Equivalent circuit diagram for pyroelectric detection: (a) the pyroelectric transducer; (b) detection electronics (Coufal et al., 1987).

1985)

$$V(t) = \frac{pA \exp(-t/\tau)}{C} \int_0^t \exp(t'/\tau) \left\langle \frac{dT}{dt'} \right\rangle dt' \qquad (8.3)$$

where t is time; $C = C_P + C_E$, with C_P being the pyroelectric capacitance and C_E the input capacitance of the preamplifier; τ is a characteristic circuit time constant, $\tau = R_T C$; and

$$R_T = \left(\frac{1}{R_P} + \frac{1}{R_E} \right)^{-1} \qquad (8.4)$$

where R_P and R_E are the resistance of the pyroelectric capacitor and of the preamplifier input, respectively. If the temperature varies sinusoidally at frequency f, the pyroelectric voltage will be

$$V(f) = \frac{2\pi i f p R_T A \langle T \rangle}{1 + 2\pi i f \tau} \qquad (8.5)$$

The temperature distribution across the pyroelectric transducer (z-direction) may be estimated using the Fourier heat conduction equation

$$\frac{\partial T}{\partial t} = D_T \frac{\partial^2 T}{\partial z^2} \qquad (8.6)$$

where D_T is the thermal diffusivity of the pyroelectric detector, defined as the ratio of the thermal conductivity k to the volume heat capacity C_v. Thus, it is easy to show that

$$\left\langle \frac{\partial T}{\partial t} \right\rangle = \frac{k}{C_v} \left\langle \frac{\partial^2 T}{\partial z^2} \right\rangle = \frac{1}{C_v z_0} \Delta H(t) \qquad (8.7)$$

where $\Delta H(t) = H(0, t) - H(z_0, t)$ is the net heat flux into the pyroelectric transducer of thickness z_0. Substituting Eq. (8.7) into Eq. (8.3) leads to

$$V(t) = \frac{pA \exp(-t/\tau)}{CC_v z_0} \int_0^t \Delta H(t') \exp(t'/\tau) \, dt' \qquad (8.8)$$

Evaluating the pyroelectric response at $t = \tau_{\text{th}}$, where τ_{th} is a thermal time constant which is long compared to the electrical time constant τ of the pyroelectric, one finds that Eq. (8.8) becomes

$$V(t) = \frac{pR_T A}{C_v z_0} \Delta H(t) \qquad (8.9)$$

In Table 8.1 some typical experimental parameters can be found for the ac and dc devices in the second and third columns, respectively. In the following discussion we will use Eq. (8.9) as the condition $\tau_{\text{th}} \gg \tau$ is satisfied in the case of available experimental work.

Table 8.1. Some Experimental Parameters Useful for dc and ac Pyroelectric Hydrogen Detectors

Experimental Parameters	Pd–PPE[a]	Pd–LiTaO$_3$[b]
Pyroelectric film thickness, z_0 (cm)	2.8×10^{-3}	23×10^{-3}
Amplifier input resistance at 20 Hz, $R_A (\Omega)$	10^8	10^9
Amplifier input capacitance, C_A (pF)	16	20
Pd thickness, L (cm)	2.85×10^{-6}	—
Pd total Area, A (cm^{-2})	0.1	0.1

[a]Christofides and Mandelis (1989a).
[b]Data from Zemel (1985).

8.3. THE dc PYROELECTRIC CHIP: Pd–LiTaO₃

8.3.1. Experimental Apparatus

A schematic diagram of a dc pyroelectric gas analyzer (dc PGA) is shown in Fig. 8.4(a). The dc PGA is made of Z-cut, single-crystal LiTaO₃ wafers, typically 230 μm thick. Fabrication steps for the device are basically the same as those used in semiconductor device photolithography: the electrodes (NiCr) are fabricated photolithographically. One of the NiCr surfaces has been coated with a catalyst material (Pd in this case). It is well known that both palladium (Pd) and nickel (Ni) have high hydrogen solubilities. At room temperature, however, H_2 is more than a thousand times more soluble in Pd than in Ni (Lewis, 1967). According to the aforementioned author, palladium has been preferred as a filter for hydrogen purification because of its selectivity to hydrogen absorption and has also been used to provide hydrogen selectivity for several types of hydrogen detectors. A heater was integrated in the pyroelectric device in order to control the temperature–time profile, $T(t)$. The differential voltage, ΔV, of PGA is directly proportional to the net heat flux into the pyroelectric device and is given by the relation

$$\Delta V = V_{Pd} - V_{IN} \qquad (8.10)$$

where V_{Pd} is the signal obtained by the Pd–NiCr electrode, and V_{IN} is the signal obtained by the uncoated inactive reference electrode. A heater was incorporated into the pyroelectric LiTaO₃ layer [Fig. 8.4(b)]. Figure 8.4(c) is another schematic drawing of a differential pyroelectric microcalorimeter. In this illustration, one of the heater leads is grounded and serves as the ground plane for the system. The transducer thus absorbs the thermal energy produced by the heater and gives rise to two different electrical voltages: V_{Pd} and V_{IN} [see Fig. 8.4(b)]. From Eqs. (8.9) and (8.10) the differential output signal may be written as (Zemel, 1985)

$$\Delta V = \frac{pAR_T}{C_v z_0}(\Delta H_{Pd} - \Delta H_{IN}) \qquad (8.11)$$

where ΔH_{Pd} and ΔH_{IN} are two net heat fluxes in the two electrodes Pd–NiCr and inactive reference–NiCr, respectively. Since the heat flow in both pyro-electric materials from the heater is the same [see Fig. 8.4(b)], the differential voltage is zero. As the reference electrode contains no chemically sensitive catalyst, its heat loss is the reference state for the measurements. On the other hand, any additional heat gain or loss depends on whether the chemical reaction is exothermic or not. The sign of the differential ΔV response will be an indication of the kind of interaction associated with the chemical processes on the catalyst. The introduction of hydrogen into the test cell and subsequent absorption of the gas by Pd cause extra heat δH in the Pd–NiCr electrode, and thus the differential voltage is

$$\delta V = \frac{A p R_T}{z_0 C_v} \delta H \qquad (8.12)$$

The adsorption and desorption of hydrogen, due to changes in its partial

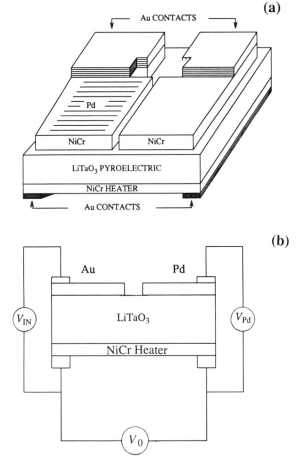

Figure 8.4. (a) Schematic of the dc pyroelectric detector showing details of construction. (b) Side view of the dc pyroelectric detector (Zemel, 1985). (c) Schematic drawing of a differential pyroelectric microcalorimeter (Zemel, 1990).

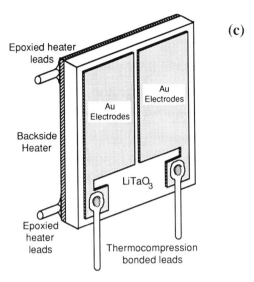

(c)

Figure 8.4 (*Continued*)

pressure in the cell environment, generates δH and thus a temperature gradient between the two electrodes.

8.3.2. Ultimate Concentration Sensitivity

From Eq. (8.12), if the detectable minimum voltage is on the order of 100 μV and the R_T is on the order of 100 MΩ, according to Zemel (1990) the detectable energy flux is on the order of 4×10^{-6} W/cm². The energy corresponds to a hydrogen detectable limit of approximately 2.45×10^9 molecules/cm³, which is less than 1 ppm of H_2 in air. In fact the number of molecules, n_{H_2}, can be determined by calculating the ratio of the minimum detectable energy flux, E_{min}, and the energy of adsorption, E_{ads}:

$$n_{H_2} = \frac{E_{min}}{E_{ads}} \approx 2.45 \times 10^9 \quad [\text{molecules/cm}^3] \qquad (8.13)$$

8.3.3. Results and Discussion

The dc PGA was exposed to a mixture of 1% hydrogen in nitrogen flowing at atmospheric pressure over the device at a rate of 180 mL/min. Typical experimental results are given in Fig. 8.5(a, b), showing the variation of the reference signal V_{IN} and the Pd–NiCr signal V_{Pd} as a function of time, separately. In Fig. 8.5(a, b) one can note the influence of the introduction

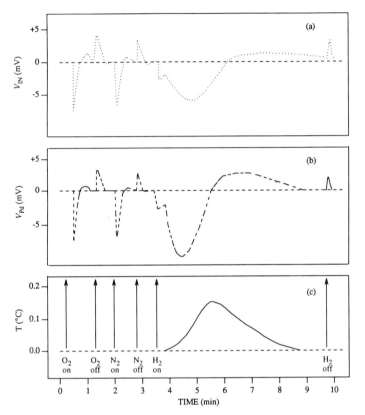

Figure 8.5. Individual component responses of the dc PGA to the admission of various gases in the measurement chamber: (a) the Pd electrode; (b) the gold electrode; (c) the temperature profile (Zemel, 1985).

(and cutoff) of O_2, N_2, and H_2 on the Pd–PGA through the variation of V_{IN} and V_{Pd} signals. According to Zemel (1985), the negative portion arises from the cooling of the dc pyroelectric detector after the reactions have reached saturation. Figure 8.5(c) shows the temperature profile of the Pd–PGA. We note that the temperature varies only upon introduction of hydrogen. From Fig. 8.5(c) we can conclude that the smaller signals (after O_2-on and -off and N_2-on and -off) may be due to external thermal disturbances when valves are opened or closed during the gas cycles. Figure 8.6 shows the variation of the differential PGA signal δV as a function of time [data collected from Fig. 8.5(a, b)]. The above preliminary results are only qualitative.

D'Amico and Zemel (1985) have also used the $Pd/LiTaO_3/Au$ structure to study the H_2/Pd interaction on Pd layers. They proved the sensitivity of

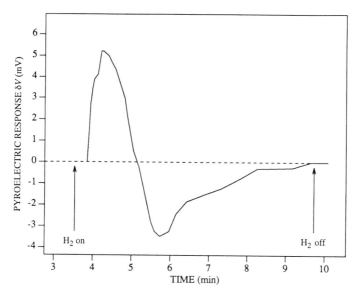

Figure 8.6. PGA pyroelectric response as a function of time ($[H_2] = 1\%$) [data collected from Fig. 8.5(a, b)].

these devices for hydrogen concentrations of 1000 ppm in nitrogen carrier gas at temperatures of 20, 55, 110, and 130 °C and made the following important observations:

a. When the working temperature, flow rate, and hydrogen concentration increase, the time response of the PE signal decreases, which indicates that the mechanisms of the catalytic process speed up.

b. The heat developed after the exposure of Pd to hydrogen depends strongly on the working temperature; the signal increases more than two times when the temperature changes from 20 to 130 °C.

c. When the hydrogen concentration increases, the heat of desorption also increases.

Finally, we should point out that D'Amico and Zemel (1985) used the *PE* device (Pd/LiTaO₃/Au) as a pyroelectric enthalpimeter. These authors inserted the PGA detector in a chamber at pressures slightly above atmospheric and proved the ability of the device to detect $[H_2]$ down to 100 ppm in a pure nitrogen flow. D'Amico and Zemel (1985) further introduced the idea of using the PGA device as a highly sensitive instrument for analyzing the kinetic response of gases and liquids on catalytic surfaces.

Besides its multilayer fabrication complexity, this dc pyroelectric sensor has proved to be susceptible to temperature fluctuations due to environmental factors (e.g. valve opening and closing during gas cycles). An additional disadvantage of this (otherwise quite sensitive) sensor may be the long delay for gas detection due to diffusive transport of thermal energy from the Pd electrode region to the reference (Au) electrode. These factors appear to have limited the development of purely pyroelectric devices as gas sensors to date.

8.4.   THE dc PYROELECTRIC CHIP: Pt–PZFNTU–Au

8.4.1.   Experimental Apparatus

A pyroelectric gas sensor has also been made by using a pyroelectric element made of PZFNTU [$Pb(Zr_{1-2x-y}Fe_xNb_xTi_y)_{1-z}U_zO_3$] ceramic (Whatmore and Bell, 1981). The arrangement of this device is shown in Fig. 8.7 (Hall et al., 1984). A dual differential configuration with the element back to back was used. The dimension of each pyroelectric PZFNTU element was 0.3 mm × 0.2 mm × 0.03 mm. The reference electrode was coated with gold, while the active element was coated with black platinum. A JEET model (BF800) was used to amplify the PE voltage signal.

8.4.2.   Results and Sensor Performance

According to Hall et al. (1984), the catalytic oxidation of hydrogen gas on black platinum at room temperature can be described by

$$H_2 + \tfrac{1}{2}O_2 \rightarrow H_2O - \delta H \qquad (8.14)$$

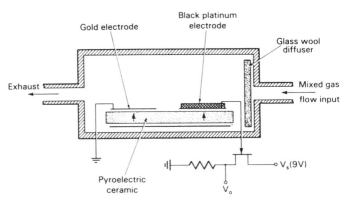

Figure 8.7. Schematic of the dc pyroelectric Pt/PZFNTU/Au gas detector (Hall et al., 1984).

where $\delta H \approx 286\,\text{kJ}\cdot\text{mol}^{-1}$. Hall et al. (1984) have performed some experiments pertaining to the detection of hydrogen in air carrier gas and the detection of oxygen in hydrogen/nitrogen mixtures. Hall et al. (1984) have performed their measurements under flow conditions at room temperature. Their system design was offering the possibility for simultaneous introduction of oxygen/nitrogen/hydrogen; nitrogen/hydrogen; and propane/nitrogen/oxygen mixtures. Figure 8.8(a) shows the PE detection of hydrogen in air. One can note that with the introduction of H_2 in the cell the PE voltage first increases drastically and then becomes slightly negative. Only 5 s are needed for the device to reach the saturation level and another 5 s to return to its initial (preexposure) position following the cutoff of the reactant supply to the test chamber. The PE signal given in Fig. 8.8(a) can be interpreted and the pattern of temperature changes in the device deduced by integrating the pyroelectric signal over time, as shown in Fig. 8.8(b). Hall et al. (1984, p. 213) interpreted their results in Fig. 8.8(a) as follows:

The large peak V_A corresponds to a rapid increase in element temperature with time as the reaction is initiated and proceeds at a constant rate. As an equilibrium state is reached, dT/dt falls to zero, and V_0 falls correspondingly. Interestingly, V_0 does not go to zero, but goes slightly negative indicating a slow cooling of the device as the reaction proceeds from A to B in Fig. [8.8(a)]. This may well be due to the increased rate of cooling of the device because of the higher thermal conductivity of the hydrogen.

No further work has been published toward the verification of the foregoing mechanism.

Finally it was also demonstrated by the aforementioned authors that the

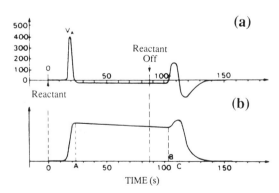

Figure 8.8. (a) PE voltage output from sensor with air as carrier gas (50 cm^3/min) and 10% hydrogen/nitrogen (50 cm^3/min) introduced as reactant at 0 s. (b) Differential temperature of detector element deduced from part (a) (Hall et al., 1984).

Pt–PE device is very sensitive: it was found to detect as low as 800 ppm of oxygen in 10% H_2/N_2 (flow rate: 1 L/min) and 300 ppm of hydrogen in air.

8.5. THE ac PHOTOPYROELECTRIC THIN-FILM DETECTOR: Pd–PVDF–Al

A few years ago, the present authors undertook the task of developing and optimizing a fast, sensitive, and inexpensive photopyroelectric (PPE) sensor for trace hydrogen gas detection at room temperature (Mandelis and Christofides, 1990a, b; Christofides and Mandelis, 1989a, b). Initially, the objective was to discover and prove the hydrogen sensing ability of thin pyroelectric poly(vinylidene fluoride) (PVDF) polymer films under gas flow-through experimental conditions in order to detect hydrogen in nitrogen ambient. In an extensive review paper, Christofides and Mandelis (1990) presented a critical comparison between the new PPE hydrogen sensor and other conventional devices. Since then, significant progress has been made, both in development and optimization of the new device (Christofides and Mandelis, 1991) and toward an understanding of the relevant physical mechanisms (Mandelis and Christofides, 1991).

8.5.1. Experimental Apparatus and Details

Poled PVDF thin films (β-phase) are known to exhibit strong pyroelectricity, i.e., a potential difference is generated in the direction of poling between the two metallized electrode surfaces that sandwich the pyroelectric film, when a temperature change is induced within the pyroelectric layer (KYNAR, 1983). The design of a hydrogen sensor fabricated from such a pyroelectric thin film has become feasible owing to the possibility of depositing a variety of thin metal electrode coatings on PVDF (Table 8.2). In the work presented by Mandelis and Christofides (1990a, b) and Christofides and Mandelis (1989a, b), PVDF was coated with Pd metal, which can adsorb and subsequently absorb hydrogen gas molecules preferentially in the presence of other ambient gases. The hypothesis was that, upon establishing an ac steady state temperature field within the pyroelectric by amplitude-modulated laser irradiation, any changes in the temperature distribution and/or the pyroelectric coefficient of the PVDF owing to interactions with the hydrogen gas would be registered as changes in the observed PPE signal, thus yielding a hydrogen sensor.

The experimental system that was used tested the response of the Pd/PVDF/Al photopyroelectric sensor to flows of hydrogen, oxygen, nitrogen, and hydrogen/oxygen mixtures, with hydrogen concentrations ranging from

Table 8.2. Some Properties of PVDF Pyroelectric Film

Properties of PVDF	
Dielectric permittivity, ϵ (F/cm)	1.06×10^{-12}
Relative permittivity, ϵ/ϵ_0	12
Capacitance, C_p (pF)	37.9
PE coefficient (at 20 °C), p_e (C/cm^2·K)	3×10^{-9}
Volume resistivity (Ω·cm)	1.5×10^{15}
Density (g/cm^3)	1.78
Thermal diffusivity, D_R (cm^2/s)	5.2×10^{-4}
Thermal time constant τ_{th} (s)	5

Source: KYNAR (1983).

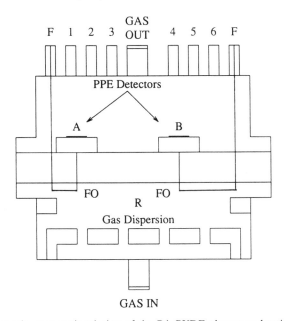

Figure 8.9. Schematic cross-sectional view of the Pd–PVDF photopyroelectric test cell: 1–6, electrical BNC connectors; F, fiber-optic adaptors; FO, fiber-optic waveguide, R, internal heating element; A&B, PPE detectors (active and reference) BNC, a trade name—originally Bayonette Non-threaded Connector.

the ppm range to pure hydrogen. The system consisted of four subsystems: a gas control subsystem; an external temperature control and test cell (see Fig. 8.9) (internal) temperature control subsystem; a fiber-optic and laser subsystem; and a signal generation and analysis subsystem. The Pd–PVDF film was placed in a standard Inficon™ housing (Coufal et al., 1987). The

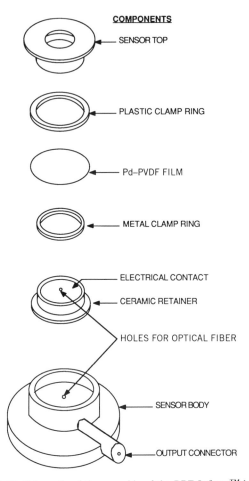

Figure 8.10. Schematic of the assembly of the PPE Inficon™ housing.

multilayer assembly, consisting of the coated PVDF pyroelectric film, was enclosed in an Inficon housing to extract the electrical signal and to ensure shielding from rf interference. The details of the sample mounting can be gathered from Fig. 8.10. The Inficon housing was placed in the experimental test cell.

An overview of the experimental setup of the PPE sensor is shown in Fig. 8.11(a). An optical source served to produce alternating temperature gradients on the Pd–PVDF and on reference Al–Ni–PVDF films, which in turn generated ac voltages due to the PPE effect. Exposure to hydrogen gas was subsequently shown to produce a controlled differential signal between

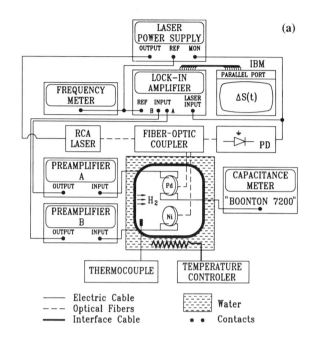

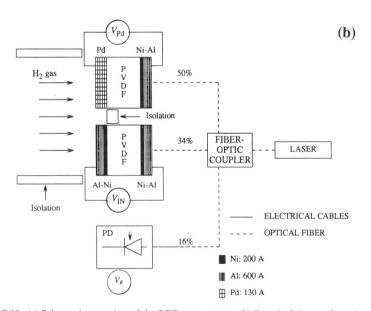

Figure 8.11. (a) Schematic overview of the PPE sensor setup. (b) Detail of the gas flow chamber, including the active Pd-coated and reference Al/Ni-coated PVDF cells (Christofides et al., 1991).

the Pd and reference electrodes; this was attributed to the adsorption, absorption, and dissociation of hydrogen molecules on the Pd surface. The signal generation and analysis of the PPE detector is also described in Fig. 8.11(a, b). The instrumentation consisted of an infrared (IR) laser diode powered by an ac current supply. The intensity-modulated output laser beam was directed to a three-way model fiber-optic coupler, where it was split into three unequal parts. The chosen optical fibers can operate even at high temperatures up to 125 °C (Szarka, 1988). One of the three beams was directed to a photodiode (PD), the output of which was then sent to the "monitor" input of the home-made laser current supply for preamplification and feedback control of the laser current, as well as to a miniaturized special-purpose lock-in amplifier capable of demodulating three signals, V_{Pd}, V_{IN}, and V_R, and outputting ΔS_R. In Fig. 8.12 one can see a schematic diagram of this home-made miniaturized differential lock-in analyzer. The feedback control consisted of using the optical reference signal to correct for temporal intensity variations in the modulated laser beam. The modulated frequency of the IR beam was monitored with a frequency meter. It is important to note that the optical absorptance of the metallized PVDF film in the IR region of the spectrum is very high (80–95%) (KYNAR, 1983), so that sensor operation was in the PPE saturation regime (Mandelis and Zver, 1985), independent of the optical properties of the coated PVDF. The two PPE signals were then connected to the differential lock-in analyzer following two low-noise Ithaco (model 1201) bandpass-filtered preamplifiers referenced by the ac laser current supply. The preamplifier gains were adjusted to minimize the lock-in output signals at the beginning of the experiments. Use of the lock-in amplifier technique in signal filtering and normalization enables the acquisition of stable differential output signals by adjusting the two gains

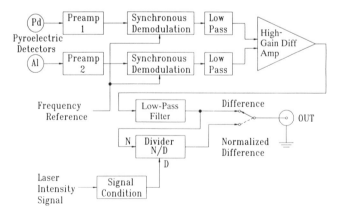

Figure 8.12. Schematic diagram of the home-made miniaturized differential lock-in analyzer (Christofides et al., 1991).

between 10 to 10,000. Providing that the input signals from the preamplifier are set to about the 10 V level, it is possible to achieve $\Delta S_R \approx 10^{-8}$, where

$$\Delta S_R = \frac{V_{Pd} - V_{IN}}{V_R} \tag{8.15}$$

For temperature control, a long heat exchanger has been used to ensure that the temperature of the incoming gas flow will be the same as that of the sensor itself. A thermocouple allowing the monitoring of the cell temperature during the experiments with a precision of $\pm 0.2\,°C$ has been used. An automatic-control heater (Fenal model 140) equipped with a second thermocouple has also been used for temperature stabilization in the range of 20–60 °C. Lower temperatures have been attained using dry ice ($-63\,°C$) and liquid nitrogen (77 K).

8.5.2. Analysis of the Photopyroelectric Response

In these experiments, harmonic modulation of the absorbed beam intensity resulted in harmonic changes in ΔT that subsequently gave rise to synchronous ac voltages $V_A(f)$ and $V_B(f)$ (Mandelis and Christofides, 1990a, b; Christofides and Mandelis, 1989a, b):

$$V_{Pd}(f) = \frac{\eta p_{Pd} W_{Pd}(f) E_{Pd_L}}{\epsilon C_{Pd} E_{Pd_{el}}} + \Phi_{Pd} \tag{8.16}$$

$$V_{IN}(f) = \frac{\eta p_{IN} W_{IN}(f) E_{IN_L}}{\epsilon C_{IN} E_{IN_{el}}} + \Phi_{IN} \tag{8.17}$$

where p_{Pd} and p_{IN} are the pyroelectric coefficients; W_{Pd} and W_{IN} are the incident optical fluxes; C_{Pd} and C_{IN} are volume specific heats; η is the optical-to-thermal energy conversion coefficient; ϵ is the PVDF dielectric constant (see Table 8.2); E_{Pd_L} and E_{IN_L} are the illuminated areas; $E_{Pd_{el}}$ and $E_{IN_{el}}$ are the electrode areas ($E_{Pd_L} < E_{Pd_{el}}$ and $E_{IN_L} < E_{IN_{el}}$); and Φ_{Pd} and Φ_{IN} are the work functions of the Pd and Al metals, respectively. For the Ni–Al metallized PVDF in the IR region of the spectrum, $\eta \approx 1$. From Eqs. (8.15)–(8.17), the output differential signal prior to the introduction of hydrogen gas in the test cell generated in the Pd–PVDF and the Al–Ni–PVDF electrodes may be shown to be

$$\Delta S(f) = \frac{1}{V_R} \left\{ \frac{\eta}{C\epsilon} \left[\frac{p_{Pd} W_{Pd}(f) E_{Pd_L}}{E_{Pd_{el}}} - \frac{p_{IN} W_{IN}(f) E_{IN_L}}{E_{IN_{el}}} \right] + [\Phi_{Pd} - \Phi_{IN}] \right\} \tag{8.18}$$

where $f = 20 \, Hz$ and $C_{Pd} = C_{IN} \equiv C$, owing to the impermeability of the PVDF film bulk by hydrogen.

Since the heat flows into the PPE devices from the laser beams were not equal $[W_{Pd} \neq W_{IN}$; see Fig. 8.11(b)], the two photopyroelectric voltages were different. Thus, the magnitude of ΔS was minimized in the beginning of each experiment by a judicious choice of gain on the preamplifiers, prior to introduction of the gas into the test cell $(V_{IN} \approx V_{Pd} \rightarrow |V_{IN} - V_{Pd}| \rightarrow \Delta S \approx 0)$. As a result of the above considerations the normalized differential voltage is given by the relation:

$$\delta S(f, [H]) = \Psi \left[\frac{p_{Pd} W_{Pd}(f) E_{Pd_L}}{E_{Pd_{el}}} - \frac{p_{IN} W_{IN}(f) E_{IN_L}}{E_{IN_{el}}} \right] + \frac{\Delta \Phi[H]}{V_R} \qquad (8.19)$$

where Ψ is an instrumental constant given by $\eta/V_R \epsilon C$ and $\Delta \Phi[H] = \Phi_{Pd} - \Phi_{IN}$. The introduction of hydrogen into the test cell does not change the heat fluxes because $\Delta T(f)$ is solely determined by the IR laser beams. From the above equality one can conclude that the introduction of hydrogen in the test cell gives rise to a photopyroelectric response because of the change of the pyroelectric coefficient *and/or* the change of the effective palladium electrode area (change in the PVDF capacitance) *and/or* a work function change. In order to identify the most probable mechanism generating the hydrogen response, both possibilities have been examined separately. A discussion concerning the fundamental mechanism of the device is given below in Section 8.7.

8.5.3. Experimental Results: Hydrogen Detection in Nitrogen

Typical experimental results for various hydrogen concentrations are presented in Fig. 8.13(a, b), showing photopyroelectric responses as a function of time (Christofides and Mandelis, 1989a). These experiments have been performed under the same conditions ($T = 20\,°C$; flow rate: 500 mL/min) and Pd coatings of two different thicknesses. At the times indicated by the H_2-on markers, the sample gas was allowed through the test cell. After saturation (horizontal part of the $\delta S(t)$ curves; t is the time) and at times indicated by the H_2-off markers, pure nitrogen flowed continuously through the cell in order to remove the absorbed hydrogen. As a result, δS decreased to the baseline existing prior to the introduction of the sample gas. Figure 8.14 shows the variation of δS_s as a function of hydrogen concentration in the range of 40–2000 ppm.

Figure 8.15 shows the variation of δS as a function of time for different H_2 flow rates (500 to 60 mL/min), $T = 20\,°C$, and H_2 concentration $= 5\%$ in

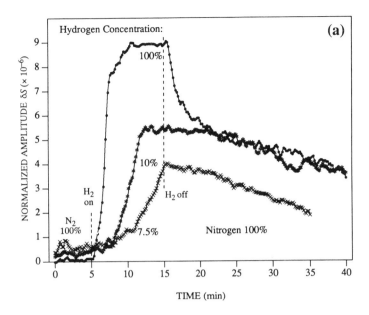

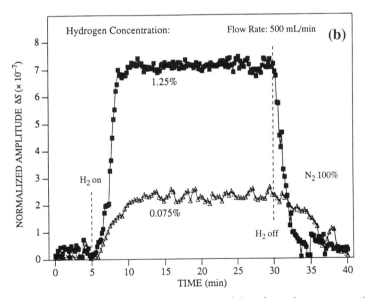

Figure 8.13. Photopyroelectric response as a function of time, for various concentrations (Pd thickness = 285 Å): (a) 7.5–100% hydrogen; (b) 1.25% and 0.075% hydrogen flow rates: 500 mL/min ($T = 20\,°C$) (Christofides and Mandelis, 1989a).

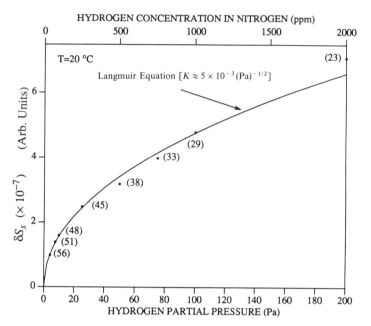

Figure 8.14. Variation of output differential saturated signal amplitude δS_s as a function of hydrogen concentration and Langmuirian isotherm fit to the data, as per Eq. (8.20) ($T = 20\,°C$; palladium thickness: 140 Å). Numbers in parentheses: response time at shown hydrogen partial pressure in minutes (Christofides and Mandelis, 1989a).

nitrogen. The flow rate was found not to influence δS_s significantly. On the other hand, it influences the response time of the sensor, which increases monotonically (from 3 to 10 min) with decreasing flow rate between 500 and 60 mL/min. The experimental results show that the PPE sensor is completely reversible and durable (Christofides and Mandelis, 1989a; Mandelis and Christofides, 1990a).

8.5.3.1. Langmuirian Analysis in the Low-Concentration Regime ($< 2000\,ppm$)

As was shown in Chapter 2, in an inert atmosphere the dissociation and association of hydrogen on a catalyst surface (palladium in the present case) may follow a number of chemical reaction steps. Figure 8.16 is a schematic· picture of possible reactions of H_2 on the Pd electrode.

In discussing a Langmuirian isotherm analysis of hydrogen coverage on Pd, in Chapters 2 and 3 we noted that the Pd bulk effect was neglected (some experimental results at very high pressures have shown that the PPE response

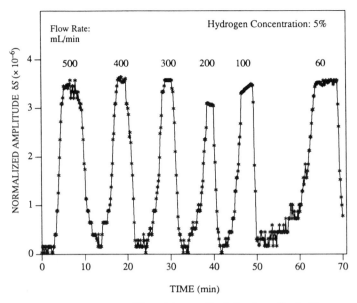

TIME (min)

Figure 8.15. (a) Variation of output differential normalized signal amplitude δS_s response delay for various hydrogen flow rates $(60-500 \, \text{mL/min}; \, T = 20 \, ^\circ\text{C})$ (Christofides and Mandelis, 1989a).

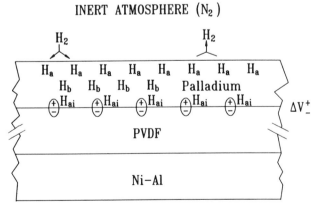

Figure 8.16. Schematic picture showing chemical reactions on the Pd surface and hydrogen transport to the bulk and to the Pd–PVDF interface (Christofides and Mandelis, 1990).

is independent of the Pd thickness). In light of results presented in a review paper by Lundström et al. (1989) concerning their Pd–MOS device and because the Pd–PPE–PVDF sensor analysis concerns very low partial pressures (< 2000 ppm), the foregoing assumption is reasonable. The device sensitivity to H_2, however, has also been attributed to a controversial Pd bulk effect (Steele et al., 1986), which is incorrect according to Lundström et al. (1989). On the other hand, according to Fortunato et al. (1989), the assumption of neglecting bulk effects at high temperature ($120\,^{\circ}$C) is not valid for room-temperature operation in the case of a Pd-gate hydrogen detector.

From the general equation of the Langmuir isotherm adapted to the Pd–PVDF experimental conditions leading to the use of Eqs. (2.10)–(2.12) we can write

$$\frac{\Theta_{is}}{1 - \Theta_{is}} = \frac{\delta S_s/\delta S_{max}}{1 + (\delta S_s/\delta S_{max})} = K(T)[p(H_2)]^{1/2} \qquad (8.20)$$

Figure 8.17(a) shows the variation of $\Theta_{is}/(1 - \Theta_{is})$ as a function of hydrogen concentration (or partial pressure) in the range of 40–2000 ppm (4–200 Pa) of hydrogen in nitrogen using a Pd film of 130 Å thickness. This figure was the main tool for the ensuing semiquantitative analysis based on surface and interface adsorption phenomena. Upon plotting $(\delta S_s/\delta S_{max})/[1 + (\delta S_s/\delta S_{max})]$ vs. $[p(H_2)]^{1/2}$ [see Fig. 8.17(a)], Eq. (8.20) allows $K(T)$ to be estimated from the slope of the curve (at $20\,^{\circ}$C). We note that the curve is linear, and from the slope we obtain $K(20\,^{\circ}$C$) \approx 4.7 \times 10^{-3}$ (Pa)$^{-1/2}$. The linearity of $[\Theta_{is}/(1 - \Theta_{is})]$ vs. $[p(H_2)]^{1/2}$ in the range of concentrations 40–2000 ppm is consistent with Lundström's (1981) findings using the Pd–MOS device. Langmuirian behavior at or about room temperature has already been reported by several authors: Hughes et al. (1987) have reported Pd–thin SiO_2–Si diodes operating at $30\,^{\circ}$C, the H_2 response of which follows a $[p(H_2)]^{1/2}$ law up to 2000 ppm (200 Pa); Fortunato et al. (1989) have also reported Langmuirian behavior up to 2000 Pa (2×10^4 ppm) at room temperature. Equation (8.20) can further be rearranged to give

$$\frac{1}{\delta S_s} - \frac{1}{\delta S_{max}} = \frac{1}{\delta S_{max}} \frac{1}{K(T)[p(H_2)]^{1/2}} \qquad (8.21)$$

In order to check the experimental conformity with Eq. (8.21), $1/\delta S_s$ vs. $1/[p(H_2)]^{1/2}$ was plotted [see Fig. 8.17(b)]. From the slope of the linear curve of Fig. 8.17(b), $K(T) \approx 5.2 \times 10^{-3}(Pa)^{-1/2}$ in excellent agreement with the value determined in Fig. 8.17(a). Furthermore, the y axis intercept of Fig. 8.17(b) gave $\delta S_{max} \approx 1 \times 10^{-5}$, in good agreement with the experimental results (experimental: $\delta S_{max} \approx 9 \times 10^{-6}$). Now, in principle it is possible to determine

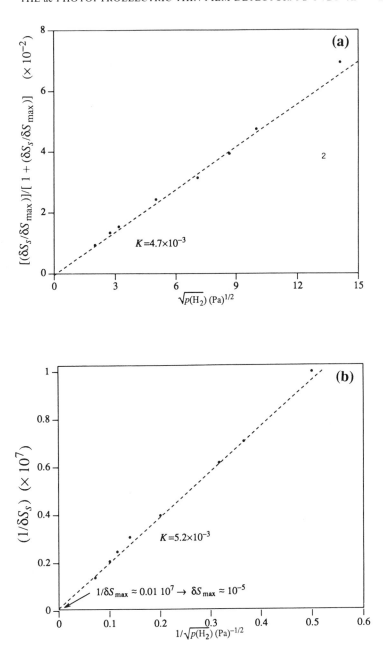

Figure 8.17. (a) $(\delta S_s/\delta S_{max})/[1 + (\delta S_s/\delta S_{max})]$ (data from Fig. 8.14 plotted as a function of square root of hydrogen partial pressure, $[p(H_2)]^{1/2}$ after Eq. (8.20). (b) $1/\delta S_s$ plotted vs. $1/[p(H_2)]^{1/2}$ after Eq. (8.21) (Christofides and Mandelis, 1989a,b, 1990).

the temperature dependence of the constant $K(T)$, and then by plotting $\ln[K(T)]$ vs. $1/T$ to obtain the value of the heat of adsorption (Lundström et al., 1989). Unfortunately, at present no such temperature-dependent measurements are available. It is easy to rewrite Eq. (8.20) in the form of the Langmuir isotherm:

$$\delta S_s = \delta S_{max}\Theta_{is} \approx \delta S_{max}\left\{\frac{K(T)[p(H_2)]^{1/2}}{1 + K(T)[p(H_2)]^{1/2}}\right\} \qquad (8.22)$$

Figure 8.14 (solid curve) shows Eq. (8.22) plotted vs. $p(H_2)$ using $K(T) = 5 \times 10^{-3}\,\text{Pa}^{-1/2}$. We note an agreement between the Langmuir isotherm and the experimental points especially at low pressures, where the Langmuirian behavior is expected to be strictly valid. Thus, the combination of the Langmuir isotherm and the PPE response makes the quantitative analysis of the signal response δS_s possible. No interpretation was given for experimental results at concentrations higher than 2000 ppm (200 Pa) because the use of the Langmuir isotherm for concentrations over 2 torr (266 Pa) is quite problematic (Dannetun et al., 1984). It turns out that a simple adsorption mechanism is not sufficient to explain and analyze the H_2 adsorption at the interface at high pressures. However, according to Lynch and Flanagan (1973), the adsorption continues after the monolayer is presumably complete. These authors have shown that, although at 37.5 °C the isotherm is almost saturated around 7 torr ($\approx 933\,\text{Pa}$), at 25 °C and at the same pressure the isotherm does not show any tendency to saturate.

8.5.3.2. Ultimate Pressure Sensitivity

Equation (8.20) predicts that with the Pd–PPE differential signal resolution ($\approx 10^{-8}$), the PPE sensor should be sensitive to a minimum coverage, Θ_{min}, at STP of approximately

$$\Theta_{min} = \frac{\delta S_{min}}{\delta S_{max}} \approx \frac{10^{-8}}{9 \times 10^{-6}} \approx 1.1 \times 10^{-3} \qquad (8.23)$$

Using Eq. (8.20) and the minimum detectable coverage, one can show that the PPE sensor should have the lowest pressure sensitivity p_{lim}:

$$p_{lim} = \left[\frac{\Theta_{min}}{K(T)(1 + \Theta_{min})}\right]^2 \approx 0.04\,\text{Pa} \qquad (8.24)$$

which corresponds to 0.4 ppm. This estimate is 100 times below the reported experimental sensitivity limit of 40 ppm of hydrogen in nitrogen (Christofides and Mandelis, 1989a). This is probably because the instrumental detection limit of 2 ppm is valid for measurements without any interferences from other gases, unlike the reported flow-through experiments. The potential sensitivity of the PPE sensor, however, to such low hydrogen partial pressures makes it a promising detection device for trace hydrogen gas analysis under STP conditions. On the other hand it is also important to note that the PPE detector is more sensitive (at room temperature) compared to some other hydrogen detectors reported by several authors, such as the Pd/a-Si:H MIS Schottky barrier diode reported by D'Amico et al. (1982). The sensitivity of the above MIS detector is 100 ppm, more than two times lower than that of the photopyroelectric sensor.

8.5.3.3. Influence of the Palladium History and Thickness on the PPE Signal

It has been discovered that the history of the Pd–PVDF active element (Pd deposition, exposure to air) is an important factor affecting performance. For example, a Pd–PVDF film exposed to laboratory air for several months becomes inactive in the presence of hydrogen, presumably owing to oxidation and molecular impurities diffused on the Pd layer (Vannice et al., 1970). Christofides and Mandelis (1989a) have also shown the influence of oxygen on the Pd surface by using a piezoelectric sensor. The sensitivity to oxygen is reversible, and the film can be reactivated after several exposures to pure hydrogen. It was found necessary, however, to keep the Pd film in a low-vacuum environment in order to protect it from various types of impurities. Other Pd-based sensors present similar long-term contamination effects and become impaired (Abe and Hosoya, 1984; Lundström et al., 1989).

The sensitivity of the Pd–PPE device depends strongly on Pd thickness. It has been shown that thick Pd layers (ca. 1588 Å) exhibit a strong response to H_2 gas (Christofides and Mandelis, 1990).

Figure 8.18 shows the variation of the normalized PPE saturated differential signal δS_s as a function of the logarithm of the Pd thickness L in the case of pure hydrogen at 20 °C. The logarithmic scale has been used only for convenience. To obtain these measurements, Pd was deposited on pyroelectric PVDF film of 28 μm thickness. In Fig. 8.18, the datum represented by a large circle well above the straight line (at $L = 130$ Å) corresponds to the only 52 μm thick film, as expected from the capacitive response. Furthermore, it was found that Pd–Cu–PVDF films respond in the same quantitative way as the Pd–PVDF films. This is reasonable, because when Cu is used as an interfacial electrical conductor between Pd and the PVDF surface, one

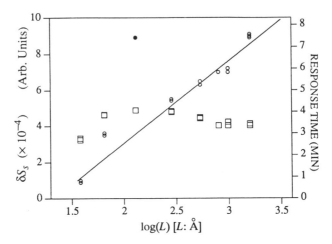

Figure 8.18. Photopyroelectric saturated signal ΔS_s (○) and response time (□) as a function of Pd thickness in the presence of pure hydrogen at room temperature. Flow rate: 500 mL/min ($T = 18\,°C$ and $f = 32\,Hz$). (●) Response of thicker PVDF, consistent with the capacitive nature of PPE signals (Christofides and Mandelis, 1989a).

expects no electric field in that layer:

$$E = -\nabla\phi(r) = 0 \qquad (8.25)$$

$$\phi(r) = \text{constant} \qquad (8.26)$$

i.e., there should be no measurable voltage drop across the electrical conductor. This result shows that the hydrogen-sensing action can be achieved upon Pd deposition, even with previously metallized PVDF, and is indicative of the fabrication flexibility built into the PPE sensor.

In terms of signal trends shown in Fig. 8.18, qualitatively similar results have been reported by D'Amico et al. (1982), who performed measurements using a palladium-coated surface acoustic wave (SAW) device. These authors studied the effect of the palladium thickness on the SAW response in a thicker range of Pd (1900–7600 Å). Buccur et al. (1976) have also studied the effect of the Pd thickness on the hydrogen/Pd interface by using a piezoelectric quartz crystal microbalance. These authors have further examined the effect of thickness on the rate of hydrogen desorption. In the case of the PPE sensor, the increase of the PPE signal with the Pd thickness is consistent with an increased proton charge density in the Pd matrix. A high-quality, fast capacitance meter (Boonton 7200) has been used for direct capacitance measurements in the presence and absence of hydrogen. These measurements have shown no measurable (i.e., < 1 pF) capacitance change. It is important to note that, as was shown by Lundström et al. (1989), their MOSFET-based device presents a capacitive response in the presence of hydrogen gas on the order of 100 pF in a hydrogen concentration of 1200 ppm. However, the detection mechanism in the MOSFET device is fundamentally different from that of the Pd–PVDF device. The capacitance insensitivity of the latter device to [H] presence shows that no depletion layer participates in the PPE response, which is consistent with the hypothesis that only virtual charge (i.e., dispacement or polarization charges) on the PVDF side of the interface may be responsible for coupling to the externally introduced charge density on the Pd side of the interface (Mandelis and Christofides, 1991).

In order to test the influence of the Pd thickness on the response time of the device, we plot in Fig. 8.18 the response time as a function of palladium thickness. There is no significant effect of the Pd thickness on the response time, which remains approximately constant and close to 3 min for every thickness and for wide thickness ranges, between 34 and 1588 Å. From the known diffusion coefficient of hydrogen in bulk Pd at room temperature ($\approx 2 \times 10^{-7}$ cm$^2 \cdot$s^{-1}) (Lewis, 1967), it follows that H atoms diffuse into the bulk over a mean distance of 1 μm of Pd in 5 s (Conrad et al., 1974; Jewett and Makrides, 1965). This estimate is consistent with Fig. 8.18, which clearly indicates that H diffusion through the Pd layer is not the rate-limiting factor in the kinetics of the Pd–PVDF sensor. Buccur et al. (1976) have also mentioned that for very small thickness (< 5000 Å) the diffusion is very rapid and it is not easy to see the difference in response time. An attempt to give a quantitative interpretation to the observed rates of adsorption–absorption–desorption based on signal response through use of the well-known Wagner (1932) model did not succeed because of the STP working conditions. Very low pressures (< 10^{-5} torr) will be needed for quantitative kinetic analysis. Comparisons with the Wagner model resulted in slower-than-predicted rates, which is indicative of the presence and influence of surface impurity molecules controlling the kinetics of the adsorption–absorption–desorption process. Specifically, the kinetic process under STP conditions takes 180 s, which is 35 times longer than the calculated value under clean surface conditions. These observations point to the importance of further experimentation under high vacuum and with molecularly clean surfaces.

8.5.3.4.   Influence of Ambient Temperature on the PPE Sensor

A very important property of the Pd–PVDF sensor proved to be its extraordinary ability to sustain unattenuated sensitivity at low temperatures continuously monitored down to $-63\,°$C and also qualitatively observed in the liquid nitrogen temperature range. No other H$_2$ gas sensor appears to have ever been reported with such ability (or, indeed, with any response at all below 20 °C) (Christofides and Mandelis, 1991; Christofides et al., 1991). As such, this property makes the sensor very promising for outdoor detection in cold climates and for fundamental studies of the hydrogen–palladium system. Temperature measurements from −63 to +50 °C have been successfully performed, indicating that good detectivity of the sensor is possible across a broad range of low temperatures. Owing to the high sensitivity of the pyroelectric effect to temperature fluctuations, special attention has been paid to obtaining high-stability background signals prior to introducing hydrogen gas to the chamber. In the absence of hydrogen gas in that temperature range, the background signal depends on the direction of temperature variation. Another important problem with low-temperature measurements (below 0 °C) was presented by the extreme difficulty in evacuating the hydrogen from the palladium. Because of this problem, low-temperature data acquisition had to be preceded by cycling up to room temperature in order to clean the surface (and bulk) of the palladium. Figure 8.19 shows the variation of the PPE saturated signal vs. temperature, which was found to satisfy a $T^{-1.4 \pm 0.15}$ law. This dependence is consistent with a thermostatistical mechanism based on hydrogenic image dipoles performing independent vibrational and librational motions about the polarization field axis of the pyroelectric (Mopsik and Broadhurst, 1975; Mandelis and Christofides, 1991). More detailed analysis is given below in Section 8.7. Figure 8.20 shows the variation of the

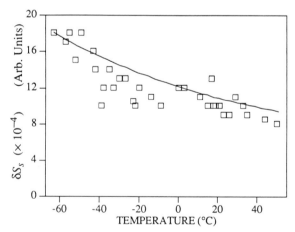

Figure 8.19. Photopyroelectric saturated signal δS_s as a function of temperature in the presence of pure hydrogen. Flow rate: 500 mL/min ($f = 32$ Hz, and $L = 1588$ Å). Solid line represents the best fit to the data ($T^{-1.4}$ law) (Christofides et al., 1991).

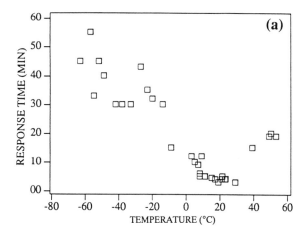

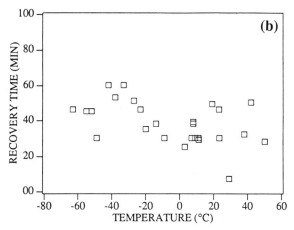

Figure 8.20. Photopyroelectric response (a) and recovery (b) times as functions of temperature in the presence of pure hydrogen. Flow rate: 500 mL/min ($f = 32$ Hz, $L = 1588$ Å) (Christofides et al., 1991).

response and recovery time as a function of the working temperature. We note that the response time of Fig. 8.20(a) depends strongly on temperature: it exhibits a minimum at ca. 20 °C and is seen to increase with decreasing temperature. No strong dependence of the recovery time on temperature can be seen in Fig. 8.20(b).

8.5.4.   Experimental Results: PPE Detection of Hydrogen/Oxygen Mixtures

It has been found that the Pd–PPE sensor can detect trace hydrogen gas in the presence of pure oxygen without significant drift and stabilization problems. The detector has been used without a reference sensor (single mode) for H_2/O_2 detection down to 0.1% of hydrogen in oxygen, which simplifies the sensor system compared to that discussed in the previous section, and reduces the cost. Here we shall describe the detection of hydrogen gas in a hydrogen/oxygen mixture. For this series of measurements the optical system consists of a solid state laser diode powered by an ac current supply as shown in Fig. 8.11. Three different signals were monitored:

$$S_{Pd} = \frac{V_{Pd}}{V_R} \qquad \Delta S_R = \frac{V_{Pd} - V_{IN}}{V_R} \qquad \Delta S_{IN} = \frac{V_{Pd} - V_{IN}}{V_{IN}} \qquad (8.27)$$

Palladium–PVDF pyroelectric films were used (Pd thickness: 1588 Å; PVDF film thickness: 52 μm). For all experiments the gas flow rate was 500 mL/min.

Before the study of the PPE response on hydrogen/oxygen mixtures, it was checked whether pure oxygen underwent any reactions with the Pd–PVDF active elements, which might introduce signal shift. Thus, initially, pure oxygen flow was introduced through the test cell containing a clean Pd–PVDF film. Figure 8.21 shows the variation of the PPE signal as a function of time in the presence of O_2, N_2, and again O_2, sequentially. No statistically significant change was observed. After this experiment the detector was exposed to different O_2/H_2 mixtures.

Typical experimental results are shown in Fig. 8.22 for two hydrogen concentrations. Before each of the experiments the detector was cleaned for 30 min with pure oxygen. Then the prescribed flow rate of hydrogen (500 mL/min) was introduced to the test cell. Figure 8.22(b) presents only two cycles (first and second) of a long experiment of approximately 7 h duration proving the complete reproducibility and reversibility of the PPE device in an oxygen/hydrogen mixture at room temperature. Similar results were obtained through cycling with various $[H_2]/[O_2]$ ratios (Christofides et al., 1993). For low-detection levels, high stabilization of the device is needed.

Figure 8.23 shows the variation of the saturated photopyroelectric signal, $(S_{Pd})_s$, as a function of hydrogen concentration. Note that $(S_{Pd})_s$ increases monotonically when the hydrogen concentration increases, while the lowest (0.1%) hydrogen concentration signal is not believed to be an absolute minimum. Differential signal detection is expected to improve this level of

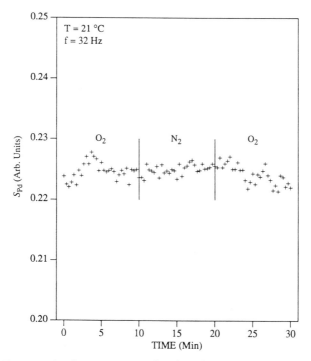

Figure 8.21. Photopyroelectric response as a function of time in the presence of pure O_2 (Christofides et al., 1993).

sensitivity. Lundström and Söderbrerg (1981/82) have succeeded in detecting much lower concentration of hydrogen in oxygen (a few ppm) by using a semiconductor MOS device. Nevertheless, there are fundamental differences between the PPE and MOSFET experimental methodologies. Lundström and Söderbrerg, as well as several other researchers in the field of hydrogen detection, performed their experiments under very specific laboratory conditions such as UHV, high temperature, clean surfaces, and static gas conditions. As the authors admitted, these conditions "are often far from those encountered when a Pd–MOS device is used as a practical sensor." On the other hand, the PPE results have been obtained from a detector relatively open to the environment, at room temperature, and without any active surface conditioning. Under these conditions, the PPE results constitute the first monitoring of hydrogen/oxygen mixtures under STP conditions.

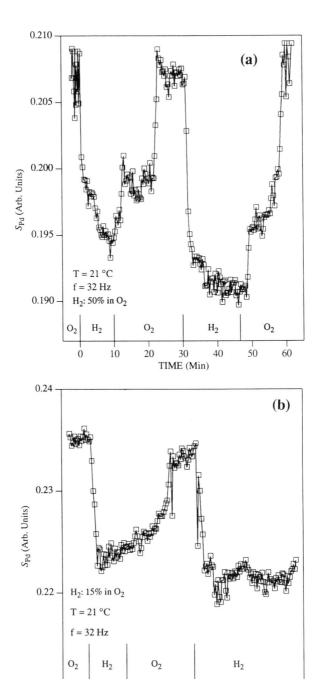

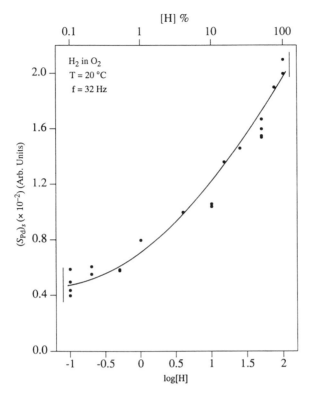

Figure 8.23. Photopyroelectric saturated signal $(S_{Pd})_s$ as a function of hydrogen concentration. The solid vertical lines represent the error at low and high concentrations (Christofides et al., 1993).

8.6. SIGNAL DRIFTS AND STABILIZATION OF THE PPE DEVICE

8.6.1. Nature of the PPE Signal Drift

As is well known, some of the main problems that appear in MOS hydrogen devices in relation to the monitoring of hydrogen concentrations are the serious signal drifts that these devices present (see Chapter 3). According to Choi et al. (1986), "despite the high sensitivity of these devices, they are still not very useful for monitoring hydrogen leaks." These nondesirable drifts

←——

Figure 8.22. Photopyroelectric signal ratio (S_{Pd}) as a function of time for two different concentrations of hydrogen oxygen: (a) 50% H_2 in O_2; (b) 15% H_2 in O_2 (Christofides et al., 1993).

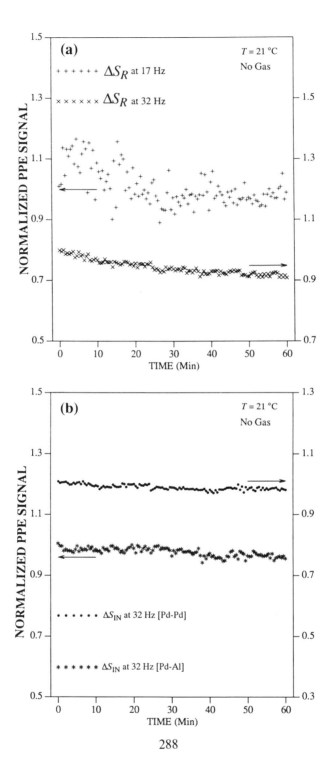

288

have been the object of several researchers in the recent past (e.g., Armgarth and Nylander, 1982; Lundström et al., 1989). According to their results, the blister formation on the Pd-gate metal is the main cause of the deterioration of sensitivity with time of the semiconductor-based devices, because they introduce a certain lattice expansion of the Pd-gate film. Another very serious signal drift of the MOS devices is the one related to charging of the oxide in the presence of hydrogen. Ultimately, it is this drift (termed HID: hydrogen-induced drift) that seriously limits the suitability of these sensors as commercial devices for continuous gas monitoring applications (Choi et al., 1986; and Chapter 3).

The drift of the PPE signal has been studied as a function of time. At time $t = 0$, the system [including the detection electronics (Fig. 8.11)] was turned on. During the first hour the signal drift was at least twice as large as that during the third hour of operation. Three main causes of signal drift could be distinguished: (a) drift related to the variation of the laser output; (b) drift associated only with the PPE background signal; and (c) drift related to the introduction of hydrogen. These have been identified as *laser drift, background drift*, and *hydrogen-induced drift*, respectively (Christofides et al., 1993).

In Fig. 8.24 signals have been normalized as shown, with the average values of 10 sampled points shown at each recorded time interval for various normalization ratios. From Fig. 8.24(a) it is seen that ΔS_R is more stable at 32 Hz than at 17 Hz, which is close to the thermally thin/thick limit (Mandelis and Zver, 1985). It is possible that the higher noise level at 17 Hz is due to the lower transfer function signal-to-noise ratio of the detection electronics (PVDF capacitor–preamplifier–lock-in analyzer), even though the absolute magnitude of the photothermal signal is substantially higher than that at 32 Hz. In fact, it has been found that the signal change in Fig. 8.24(a) is 13% and 9% at 17 and 32 Hz, respectively. On the other hand, between the two quantities ΔS_{IN} shown in Fig. 8.24(b), the one which was obtained from the two neighboring Pd–PVDF detectors of the same film is more stable, as expected, with a 3% change (vs. a 6.5% change of the Pd–Al pair) over the observation time span of Fig. 8.24(b).

8.6.2. Temperature and Background Drift

Figure 8.25 shows the variation of the PPE signal S_{Pd} as a function of time. The signal had been stabilized at $T = 20\,^{\circ}\text{C}$ before the temperature was

---

Figure 8.24. (a) Normalized PPE signal ΔS_R variation as a function of time for two frequencies and two normalization procedures using different PVDF films (see text). (b) Normalized PPE signal ΔS_{IN} variation for two different PVDF reference sensors: (i) from Pd–PVDF film adjacent to the active element ($\cdots$) (ii) from Al–Ni–PVDF film (∗∗∗). In parts (a) and (b) the signals have been artificially separated out for clarity (Christofides et al., 1993).

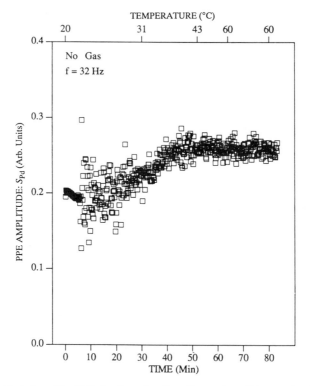

Figure 8.25. Variation of the PPE signal as a function of increasing cell temperature (Christofides et al., 1993).

ramped up, and no significant instrumental background was registered. Subsequently, the signal became very noisy at the onset of increase of the cell temperature. The PPE sensor was thus seen to need at least 1 h of stabilization before it was able to detect concentrations lower than 0.5% of H_2 in O_2 at elevated temperatures. Some preliminary results with a $LiTaO_3$ PPE sensor have shown similar relaxation times (at high temperatures) as the PVDF thin films. $LiTaO_3$ crystal pyroelectrics have a clear advantage when high temperatures ($\geqslant 100\,°C$) must be employed.

8.7. PHOTOTHERMAL ELECTROSTATICS OF THE METAL–PYROELECTRIC JUNCTION

In this section a summary description of the photothermal electrostatics of the PPE junction device is given (Mandelis and Christofides, 1991).

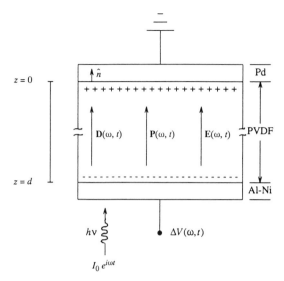

Figure 8.26. PPE H_2 capacitive sensor. The Pd anode (ground electrode) is in contact with the Inficon™ housing of the sensor. The cathode (back electrode) is the standard Pennwalt Al/Ni-layered electrode (KYNAR, 1983). Intermittent photothermal excitation of intensity I_0 at frequency $f = \omega/2\pi$ is assumed from the rear side of the capacitor (Mandelis and Christofides, 1991).

The geometry of the Pd–PVDF H_2 sensor is shown in Fig. 8.26. This amounts to a capacitor consisting of Pd-electrode (ground; anode)/PVDF dielectric/Al–Ni-electrode (cathode). For a pyroelectric dielectric such as β-PVDF with a remanent frozen-in nonequilibrium polarization **P** due to an external electric field $\mathbf{E}_{ext}$, the Gibbs free energy is

$$G(T, \mathbf{E}_{ext}) = U - TS - \mathbf{P} \cdot \mathbf{E}_{ext} \qquad (8.28)$$

where U, T, and S are the total internal energy, the temperature, and the entropy of the pyroelectric, respectively. In Eq. (8.28) the piezoelectric properties of the pyroelectric have been ignored (Zemel, 1985). From the second law of thermodynamics, the internal energy content change due to a change in polarization brought about by an external electric field is (Wangsness, 1986)

$$\Delta U = T \Delta S + \mathbf{E}_{ext} \cdot \Delta \mathbf{P} \qquad (8.29)$$

Equations (8.28) and (8.29) give the concomitant variation in the Gibbs free energy:

$$\Delta G(T, \mathbf{E}_{ext}) = -S \Delta T - \mathbf{P} \qquad (8.30)$$

Equation (8.30) helps define the polarization vector thermodynamically:

$$\mathbf{P} = \mathbf{P}(T, \mathbf{E}_{ext}) \equiv -\mathbf{V}_E G(T, \mathbf{E}_{ext})|_T \qquad (8.31)$$

where

$$\mathbf{V}_E \equiv \hat{i} \frac{\partial}{\partial E_x} + \hat{j} \frac{\partial}{\partial E_y} \hat{k} \frac{\partial}{\partial E_z} \qquad (8.32)$$

Upon consideration of the polarization change, $\Delta \mathbf{P}$, in the pyroelectric element, due to variations in the temperature and electric field, T and $\mathbf{E}_{ext}$, one further obtains from a variation in Eq. (8.31)

$$\Delta \mathbf{P}(T, \mathbf{E}_{ext}) = \left(\frac{\partial \mathbf{P}}{\partial T} \bigg|_{\mathbf{E}_{ext}} \right) \Delta T + (\mathbf{V}_E \cdot \mathbf{P}|_T) : \Delta \mathbf{E}_{ext} \qquad (8.33)$$

Equation (8.33) is important in the present development because it shows that under the most general thermodynamic equilibrium conditions a change in the polarization vector of the pyroelectric is predicted if either the temperature or the external applied electric field, *or both*, are varied. The following definitions can now be made:

$$\mathbf{p}(T) \equiv \frac{\partial}{\partial T} \mathbf{P}(T, \mathbf{E}_{ext}) \bigg|_{\mathbf{E}_{ext}} \qquad (8.34)$$

for the pyroelectric coefficient vector of the material (Broadhurst and Davis, 1987), and

$$\boldsymbol{\epsilon}(\mathbf{E}_{ext}) \equiv \mathbf{V}_E \cdot \mathbf{P}(T, \mathbf{E}_{ext})|_T \qquad (8.35)$$

for the dielectric tensor:

$$\Delta \mathbf{P}(T, \mathbf{E}_{ext}) = \mathbf{p}(T) \Delta T + \boldsymbol{\epsilon}(\mathbf{E}_{ext}) : \Delta \mathbf{E}_{ext} \qquad (8.36)$$

In the absence of an applied external field, i.e., for $\Delta \mathbf{E}_{ext} = \mathbf{0}$, a photothermal modulation of the structure in Fig. 8.26 results in synchronous oscillation of the generally depth-dependent polarization vector. From Eq. (8.36), we have

$$\Delta \mathbf{P}(\omega, z, t) = \mathbf{p}(T, z) \Delta T(\omega, z, t) \qquad (8.37)$$

This amounts to a depth-dependent oscillating internal electric field, given by (Wangsness, 1986)

$$\Delta \mathbf{E}(\omega, z, t) = \frac{1}{\epsilon_0} \chi_e^{-1} : \mathbf{p}(T, z) \Delta T(\omega, z, t) \qquad (8.38)$$

with χ_e the electric susceptibility tensor. To facilitate further analysis, without loss of the general physics of the situation, the photothermally excited pyroelectric will subsequently be assumed to behave *linearly* and *isotropically* for low incident optical irradiances I_0 (Fig. 8.26). Under these conditions both ϵ and χ_e tensors are reduced to scalars, and the potential difference $\Delta V(\omega, t)$ developed across the pyroelectric capacitor photothermally can be written as (Mandelis and Christofides, 1991)

$$\Delta V(\omega, t) = - I_0 \frac{i\eta p_z \{ 1 - \exp[- (1 + i)a_p(\omega)d] \}}{4\chi_e \epsilon_0 k_p (1 + g)a_p^2(\omega)} \exp(i\omega t) \qquad (8.39)$$

where η is the nonradiative (optical-to-thermal) energy conversion efficiency; k_p, the thermal conductivity of the pyroelectric; g, a backing-to-pyroelectric thermal coupling coefficient (Mandelis and Zver, 1985); p_z, the z component of the pyroelectric coefficient (see Fig. 8.26); and $a_p(\omega)$ is the thermal diffusion coefficient in the PVDF given by

$$a_p(\omega) = (\omega/2\alpha_p)^{1/2} \qquad (8.40)$$

Here α_p is the thermal diffusivity of the pyroelectric.

The ac photothermal potential oscillation of Eq. (8.39) modulates the energy band diagram of Fig. 8.27 (a catalytic metal–insulator–inert metal structure). In constructing the band structure of the H_2 sensor, the following were taken into account: (a) the well-known decrease (Lundström, 1981) of the Pd work function, $q\phi_{Pd}$, in the presence of a hydrogen concentration in the Pd-matrix, which forms a dipole layer at the Pd–insulator interface, as observed conclusively with the Pd–SiO$_2$ interface (Lundström et al., 1975a, b, 1977):

$$q\phi_{Pd}[H] = q(\phi_{Pd}[0] - \Delta V[H]) \qquad (8.41)$$

(b) the photothermal potential modulation of the energy levels of the pyroelectric due to the incident optical irradiance $I_0 e^{i\omega t}$, which results in an effective potential source given by Eq. (8.39); and (c) the existing contact potential difference between the anode (Pd) and inactive cathode (Al/Ni) metals:

$$q\phi_0[H] = q(\phi_{Pd}[H] - \phi_{IN}) \qquad (8.42)$$

The electrostatic potential function across the device structure, which determines the curvature of the energy-band diagram in Fig. 8.27, can be calculated as a linear superposition of the three effects outlined above. The presence of a thin unpolarized layer close to the electroded surface of the

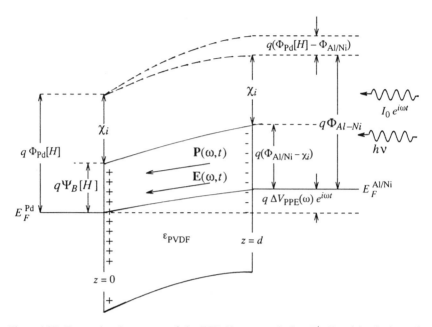

Figure 8.27. Energy band structure of the PPE H_2 sensor device: E_F^j, Fermi level of metal j ($j = Pd$, Al/Ni), χ_i, electron affinity of the insulator (PVDF film); $\phi_{Pd}[H]$, Pd work function, a function of hydrogen concentration in the catalytic metal; $\Psi_B[H]$, potential barrier height between the Pd Fermi level and the conduction band of the insulator; ϕ_{Al-Ni}, work function of the layered electrode of the anode. Intermittent photothermal excitation is assumed from the Al/Ni electrode side (Mandelis and Christofides, 1991).

pyroelectric [a few nanometers thick (Coufal et al., 1987)] has been ignorned in PPE electrostatic development. Assuming potential invariance along the lateral (x, y) dimensions, consistent with the thin-film nature of the pyroelectric PVDF, according to Mandelis and Christofides (1991) the various contributions to the potential field are as follows: (a) contact potential; (b) photopyroelectric potential; and (c) hydrogen-induced potential. The linear superposition of all the relevant potential fields, can be given by the relation

$$\Delta V(\omega, [H]; t) = \left(\frac{d}{\epsilon}\right)\sigma_f^{(PPE)}(\omega, t) + \left(\frac{d}{2\epsilon}\right)\sigma_f^{(+)}[H] + \Delta\phi_0[H] \qquad (8.43)$$

where

$$\Delta\phi_0[H] = \phi_{IN} - \phi_{Pd}[H] \qquad (8.44)$$

In synchronous lock-in operation of the device, the middle term on the right-hand side of Eq. (8.43) must be omitted, as it represents a dc potential contribution. Therefore, the observed dependence of the PPE signal on the hydrogen concentration must involve either the first term (which should carry an [H]-dependence) or the third term (which should carry an ω-dependence), or both.

8.7.1. [H]-Dependence of $\sigma_f^{(PPE)}(\omega, t)$

The presence of an unbalanced dipole layer in the vicinity of the Pd–PVDF interface produces a net change $\Delta P[H]$ in the original oscillating polarization of the pyroelectric setup by the PPE effect. An important consequence of this model is that in the limit of a photothermal modulation frequency range in which the pyroelectric is thermally thick (Mandelis and Zver, 1985) it predicts an output voltage amplitude modulation frequency dependence

$$|\Delta V_l(\omega, [H])| = \frac{I_0 \eta \alpha_p}{\epsilon(1 + g)k_p \omega} p_z[H] \propto \omega^{-1} \qquad (8.45)$$

for the thermally thick condition, and a pyroelectric coefficient, p_z, dependent on the concentration of hydrogen absorbed in the Pd matrix.

A PPE signal due to H_2 absorption could be measured up to $380\,Hz$ under continuous hydrogen flow conditions. The ratio of PPE voltage in the presence of H_2 to the voltage in ambient N_2 in Fig. 8.28 normalizes out the instrumental and geometric frequency dependences of the sensor detection system. The slopes of both voltages in Fig. 8.28 in the thermally thick regime were somewhat different from the predicted ω^{-1} dependence of Eq. (8.45), and the discrepancy was traced back to the violation of the one-dimensional assumption under which Eq. (8.45) was derived: in the reported experiments (Mandelis and Christofides, 1991) the optical fiber source size was on the order of one thermal diffusion length (a_p^{-1}), which rendered the problem three-dimensional, with an effect on the exponent of the algebraic frequency dependence of the signal. Overall, the important message of Fig. 8.28 is that the hydrogenic PPE voltage behaves consistently with a bulk pyroelectric effect, with an [H]-dependence due to the change in the pyroelectric coefficient value.

8.7.2. ω-Dependence of $\Delta\phi_0[H]$

The frequency dependence of the contact potential difference between catalytic active Pd and inert inactive metal Al/Ni electrode can be obtained upon

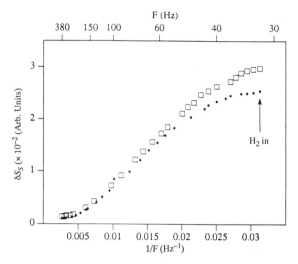

Figure 8.28. PPE saturation signal vs. frequency: (□) 100% H_2 was introduced at $f = 32\,Hz$ and frequency was increased under constant H_2 flow of 500 mL/min; (●) reference curve in 100% N_2 (Mandelis and Christofides, 1991).

consideration of the temperature dependence of the Pd work function:

$$\phi_{Pd} = \phi_{Pd}([H], T) \qquad (8.46)$$

It can be shown (Mandelis and Christofides, 1991) that the small thermal-wave modulation of the contact potential difference will produce an ac contribution to the third term on the right-hand side of Eq. (8.43) with voltage amplitude

$$|\Delta V_{III}(\omega, [H])| = -\left[\frac{\partial}{\partial T}\phi_{Pd}([H], T)\Big|_{T = T_0}\right]\Delta T_0(\omega)\exp\left[-\left(\frac{\omega}{2\alpha_p}\right)^{1/2} d\right] \qquad (8.47)$$

with

$$\Delta T_0(\omega) \approx \frac{-I_0 \eta}{2(1 + g)k_p a_p^2(\omega)} \qquad (8.48)$$

for thermally thick pyroelectric. A similar mechanism was proposed qualitatively by Balasubramanian et al. (1991) to explain the sensitivity to ambient H_2 of a Pd–mica–inert electrode (gold) capacitor. In that device there can be no PPE effect, and thus this can be the only source of a synchronous voltage dependence on [H] through the temperature derivative in the brackets of Eq. (8.47). It is important to note that the modulation frequency dependence of $|\Delta V_{III}|$ is $\approx \omega^{-1}\exp[-K(\omega^{1/2})]$ in this model, as the back-surface laser beam-induced thermomodulation of the Pd work function is only affected by the tail of the exponentially damped thermal-wave distribution within the bulk of the dielectric. Therefore, in principle, the frequency dependence of the PPE device signal, where both mechanisms expressed by Eqs. (8.45) and (8.47) are possible, can resolve which one is dominating the operation of the Pd–PPE sensor. The device signal frequency dependence under constant flow conditions exhibited essential agreement with Eq. (8.45), i.e., consistent with a change in the pyroelectric coefficient.

A separate set of experiments involved measuring the H_2 saturation voltage at each modulation frequency, then removing the hydrogen gas from the chamber, changing the frequency, and repeating the measurement. The results are shown in Fig. 8.29, which suggests an exponential decay of the saturation PPE signal with $f^{1/2}$ in the thermally thick regime. It is interesting to note that in this experimental mode the rapid decay of the differential PPE signals with increasing frequency is qualitatively, at least, in agreement with a similar observation reported at 5% H_2 in N_2 with the Pd–mica–Au sensor (Balasubramanian et al., 1991). The slopes of the two curves in Fig. 8.29 were found to give, according to Eq. (8.45), thermal diffusivities for PVDF approximately a factor of 3 (52 μm film) and 10 (28 μm film), respectively, lower than the accepted value of $5.4 \times 10^{-8}\,\text{m}^2/\text{s}$ (KYNAR, 1983). Again,

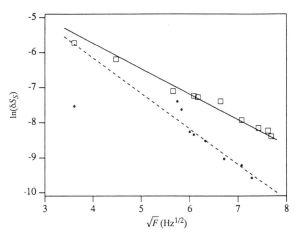

Figure 8.29. PPE saturation signal vs. frequency: 100% H_2 was introduced at each frequency at 500 mL/min and was cycled out of the system before each frequency change. Upper curve: 28 μm thick PVDF, 1588 Å thick Pd. Lower curve: 52 μm thick PVDF, 130 Å thick Pd (Mandelis and Christofides, 1991).

these low values are a result of three-dimensionality in the thermal-wave propagation problem, where the one-dimensional expression Eq. (8.47) was used for the calculation of α_p, and were also measured in earlier work with three-dimensional PPE geometries (Mieszkowski et al., 1989). The important conclusion from Fig. 8.29 is the exponential decay of the PPE voltage with $f^{1/2}$, in agreement with a Pd–surface work function temperature modulation mechanism, as advanced by Balasubramanian et al. (1991) for the Pd–mica–inert electrode system, and described by the model leading to Eq. (8.29) above. This is reasonable, as upon each new exposure to hydrogen gas it is the temperature of the Pd–surface that controls the concentration of adsorbed, dissociated, and absorbed hydrogen, while there is no residual bulk hydrogen to bias the Pd–PVDF and thus generate the bulk effect observed under continuous flow-through conditions.

Further support of the interfacial bias mechanism under continuous hydrogen flow-through conditions is provided by Fig. 8.30. Two PVDF films were coated with 285 Å Pd on oppositely polarized surfaces. It can be seen that the positively polarized electrode gives a normalized voltage, S_{Pd+}, which increases as a function of time, whereas the voltage from the negatively polarized electrode, S_{Pd-}, decreases in a manner essentially symmetric with respect to the equilibrium (i.e., $N_H = 0$) level. This behavior at a fixed modulation frequency strongly indicates the interfacial nature of the operating mechanism where the sign of one of the electrostatic charges being superposed at the Pd–PVDF interface changes (i.e., the PVDF surface charge density carries opposite signs on oppositely polarized surfaces). If the contact potential

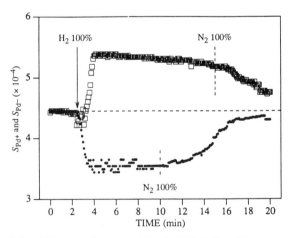

Figure 8.30. Variation of S_{Pd+} as a function of time ($\square$). Variation of S_{Pd-} as a function of time ($\bullet$) (Mandelis and Christofides, 1991).

shift were dominant, there would be no difference between the signals from oppositely polarized pyroelectrics, as long as the Pd–pyroelectric–inert metal geometry remained unaltered.

The ambient temperature dependence of the 100% H_2-saturated PPE signal was also monitored between -63 and $+50\,°C$. A slight decrease in the signal with increasing T was found, best described by the following relationship (Mandelis and Christofides, 1991):

$$\Delta V_{sat}(T) \approx \text{const.} \times T^{-1.4 \pm 0.15} \tag{8.49}$$

The temperature dependence of the PPE device behavior, reported to be measured over a broad low T (i.e., below room temperature) range for the first time with solid state H_2 sensors, offers a unique opportunity to use the Pd–PVDF junction for the study of the physics of hydrogen–Pd–polymer· interaction. Mandelis and Christofides (1991) further showed in their electro-static model that the PPE response depends on temperature as

$$\delta S \propto \left[\frac{P_n^{(PPE)}}{\chi_e \epsilon_0 \omega_v (k_B I)^{1/2}} \right] T^{-3/2} \tag{8.50}$$

which yields a $T^{-1.5}$ dependence for the pyroelectric coefficient on the temperature in agreement with experimental observations. This dependence can be obtained by assuming independent vibrational and librational motions of the pyroelectric dipoles around the axis of symmetry of the material. The decrease in the saturation PPE signal in the presence of H_2 with temperature can then be understood as being due to a thermostatistical decrease of the mean librating hydrogenic dipole moment with a temperature-induced increase in the rms fluctuation angle, and the concomitant increased libration amplitude, as discussed by Broadhurst and Davis (1987). It is interesting to note that a Pd work function thermal-wave temperature modulation mechanism, as proposed by Balasubramanian et al. (1991), leads to a T^{-2} theoretical signal dependence.

8.8. Pd–MICA–Au CAPACITIVE THERMAL-WAVE SENSOR

Balasubramanian et al. (1991) used a piece of 3 mm thick cleaved mica; 80 nm of Pd evaporated through a metal mask was used as a chemical selective layer. A reference electrode of equal size was also deposited with a gold layer in the same manner. A high-intensity light source obtained from a Xe lamp was employed. The light intensity was modulated at three frequencies: 10, 20, and 50 Hz.

As in the case of Pd semiconductor devices (Lundström et al., 1986), the change in the catalyst metal work function, ϕ_{ms}, due to exposure to gaseous H_2 is given by

$$\phi_{ms} \approx \phi_{ms,0} \frac{K(T)[p(H_2)]^{1/2}}{1 + K(T)[p(H_2)]^{1/2}} \qquad (8.51)$$

where $K(T)$ is a constant that depends on the difference in absorption energies, E_{ads}, at the gas–metal surface and the metal–substrate interface, respectively. According to Balasubramanian et al. (1991), $K(T)$ can be written as

$$K(T) = \beta_0 T^n \exp\left(-\frac{E_{ads}}{kT}\right) \qquad (8.52)$$

where β_0 in this case is a temperature-independent parameter. The total charge released in the Pd chemical layer of a pyroelectrically active device with a catalytic metal layer, Q_{Pd}, and the charge released in a reference pyroelectric device without catalytic metal, Q_{IN}, can be written, respectively, as

$$Q_{Pd} = C(V + \phi_{ms}) + p(T)(\langle T \rangle - T_0)A \qquad (8.53)$$

$$Q_{IN} = CV + p(T)(\langle T \rangle - T_0)A \qquad (8.54)$$

where A is the (common) area of the Pd and reference electrodes. In the case where light of angular modulation frequency ω impinges on the system, a periodic current of the same frequency arises:

$$I_{Pd} = i\omega\left\{CV + \left[\frac{C}{A}\frac{\partial \phi_{ms}}{\partial T} + p(T)T\frac{\partial p(T)}{\partial T}\right]A\langle T \rangle\right\} \qquad (8.55)$$

and

$$I_{IN} = i\omega\left\{CV + \left[p(T)T\frac{\partial p(T)}{\partial T}\right]A\langle T \rangle\right\} \qquad (8.56)$$

Subtracting Eq. (8.56) from Eq. (8.55) results in the differential current ΔI_{ms}, which is independent of the pyroelectric coefficient $p(T)$. The temperature dependence of ϕ_{ms} is expected to be dominated by the exponential term of Eq. (8.52)

According to Balasubramanian et al. (1991), the average temperature $\langle T \rangle$ as a function of the modulation frequency can be given by the relation

$$\langle T \rangle \approx \frac{\Delta H(0)}{MC_v(1 + i\omega\tau_T)} \qquad (8.57)$$

where τ_T is the thermal time constant of the capacitor. In the special case of low frequencies and high $p(H_2)$, the voltage shift is

$$\Delta V = \left(\frac{E_d}{kT^2} \right) \frac{\Delta H(0)\tau\phi_{ms,0}}{MC_v\tau_T} \qquad (8.58)$$

where τ is the usual RC time constant of the circuit. The important conclusion from Eq. (8.58) is that in this particular case [low frequencies and high $p(H_2)$] ΔV is independent of the modulation frequency.

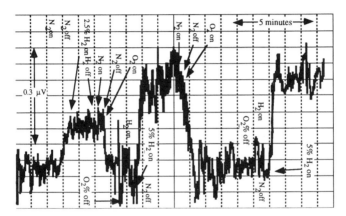

Figure 8.31. Reproduction of the time dependence of the photothermal response of a Pd–mica–Au capacitive structure to the admission of H_2 and O_2 gases (Balasubramanian et al., 1991).

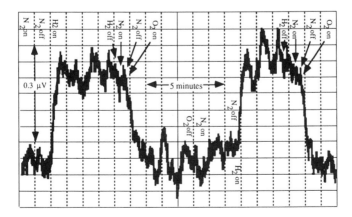

Figure 8.32. Reproduction of the time dependence of the photothermal response of a Pd–mica–Au capacitive structure to the admission of approximately 2.5% H_2, 5% H_2, and 5% O_2 gases (Balasubramanian et al., 1991).

' Figure 8.31 presents a typical reproducible response of the Pd–Mica–Au device as a function of time at two hydrogen exposures (5% of hydrogen in nitrogen). Figure 8.32 shows the ΔV shift as a function of time for hydrogen concentrations: 2.5% and 5%. The data presented on the aforementioned two figures were obtained with the modulation frequency at 20 Hz. Measurements at 10 Mz have given approximately the same voltage shift responses. Balasubramanian et al. (1991) have pointed out that no measurable signal was observed when the modulation frequency was increased to 50 Hz.

8.9. OTHER APPLICATIONS OF PYROELECTRIC SENSORS

The pyroelectric property offers the possibility to use PE materials for a number of applications. As Whatmore (1986) pointed out: "their a.c.-coupled nature makes them insensitive to unvarying fluxes of radiation so that they are ideally suited to detecting small changes in a relatively large background level of incident energy." In fact, PE sensors possess five main advantages that make them suitable for many applications:

a. They are sensitive in a very large spectral bandwidth.
b. They are sensitive in a very wide temperature range without need for cooling.
c. They have low power requirements.
d. They give a fast response.
e. They are made with generally low-cost materials.

In this section we shall briefly describe the state of the art of pyroelectric sensors in the field of instrument and measurement science. PPE detection of gases is a fairly recent phase of the application of pyroelectric devices, having started only in 1987. In the beginning, the pyroelectric effect was employed mainly for the detection of radiation and for laser power measurements, as well as in solar technology.

The pyroelectric phenomenon has led to many applications. In fact, since the beginning of the twentieth century pyroelectricity has been explored in diverse areas, especially for the detection of radiation, as just noted. In the early 1960s PE detectors were developed in the visible range (Cooper, 1962) through the IR (Putley, 1970) to submillimeter wavelengths (Hadni, 1963; Hadni et al., 1978). Putley (1970) in an excellent chapter reviewed the use of pyroelectricity for the detection of radiation. Modulated PE radiation detectors have also been developed in the range of a few hertz (Putley, 1970) to gigahertz (Roundy et al., 1974). Later, an electrostatic copying process based on the pyroelectric effect was also described (Bergman et al., 1972), and since

Table 8.3. Various Applications of Pyroelectric Detectors Other Than Gas Sensing

References	Field of PE Applications
Cooper (1962)	Radiation monitor: visible
Hadni (1963)	Radiation monitor: sub-mm
Putley (1970)	Radiation monitor: Hz
Putley (1970)	Radiation monitor: IR
Roundy et al. (1974)	Radiation monitor: GHz
Yokoo et al. (1986)	Radiation monitor
Ploss et al. (1990)	Radiation monitor
Lang (1962)	PE thermometer
Whatmore (1986)	PE intruder alarm
Coufal et al. (1987)	PE calorimeter
D'Amico and Zemel (1985)	PE enthalpimetry
Bergman et al. (1972)	PE copying
Rahnamai and Zemel (1981/82)	PE Anemometer
Rahnamai (1982)	PE Anemometer
Rahnamai and Zemel (1983)	PE Anemometer
Frederick et al. (1985)	PE Anemometer
Hesketh et al. (1985)	PE Anemometer
Watton (1976)	PE Vidicon
Okuyama et al. (1989)	PE IR–CCD[a] imaging
Baumann et al. (1983)	Thermal-wave microscopy
Stillwell (1981)	Thermal-wave imaging
Mieszkowski et al. (1989)	Thermal-wave tomography
Munidasa and Mandelis (1991)	Thermal-wave tomography
Simpson and Tuzzolino (1984)	Pulse detector of nuclei
Burns et al. (1983)	Material characterization
Coufal (1984)	Material characterization
Mandelis (1984)	Material characterization
Christofides et al. (1990)	Material characterization
Marinelli et al. (1992)	Material characterization
Faria et al. (1986)	Device characterization
Mandelis et al. (1990)	PPE–FTIR spectroscopy
Mandelis and Power (1989)	PPE impulse response measurements

[a]CCD = charge-coupled-device.

303

then the applications of pyroelectricity have become widespread: the PE thermometer (Lang, 1962), the PE anemometer (Rahnamai and Zemel, 1981/82, 1983; Rahnamai, 1982; Frederick et al., 1985), the radiation monitor (Hadni, 1980), and the PE IR image sensors (Okuyama et al., 1989), among others. In recent years, the PE sensors have also found applications in thermal-wave tomography for subsurface defect detection (Mieszkowski et al., 1989; Munidasa and Mandelis, 1991). In the 1980s PE sensors were used for characterization of advanced materials (Burns et al., 1983; Coufal, 1984; Mandelis, 1984) and devices (Faria et al., 1986). An effort toward the development of PE spectroscopic measurement instrumentation has also been made (Christofides et al., 1990; Mandelis et al., 1990; Mandelis and Power, 1989). Table 8.3 reviews various applications of pyroelectricity and presents some of the most noteworthy applications in the field.

8.10. GENERAL CONCLUSIONS

The main conclusions of this chapter can be summarized as follows:

1. The Pt–PE device (based on PZFNTU ceramic) has been found to detect as low as 800 ppm of oxygen in 10% H_2/N_2 (flow rate: 1 L/min) and 300 ppm of hydrogen in air.

2. The Pd–$LiTaO_3$–Au pyroelectric sensor has been found to be sensitive to hydrogen concentrations of 1000 ppm in nitrogen carrier gas at temperatures of 20, 55, 110, and 130 °C.

3. The ac photopyroelectric Pd–PVDF device has been found to be more sensitive than the dc pyroelectric sensor in the presence of pure nitrogen.

4. From the point of view of the physics of the Pd–PVDF–PPE detector operation, a semiquantitative phenomenological understanding has been achieved by showing that the detector response is consistent with elementary adsorption mechanisms in the Pd–H_2 system in the low H_2 partial pressure range (< 200 Pa).

5. A quantitative interpretation of the hydrogen partial pressure dependence of the differential signal has been achieved by using simple gas–solid interaction theory and the combination of the Langmuir isotherm with PPE theory.

6. It has been found that the thickness of Pd evaporated on PVDF film plays an important role in determining sensitivity and durability. Thus Pd film thickness optimization is of intense interest for the optimization

of the sensor characteristics (durability, reversibility, sensitivity, and speed of response).

7. Extensions of the reported measurements on hydrogen and oxygen gas detection to the detection of other gases may lead to an increase of the PPE detector utility. Other environmentally important gases may be detectable, such as hydrocarbons, SO_2, HCl, NH_3, H_2S, and even radioactive gases such as tritium.

8. The hydrogen PPE detector might well also become an excellent tool for surface science studies under UHV, especially at low temperatures where other sensors do not exhibit good performance. In fact, UHV studies should also help further a better understanding of the operating mechanism, because of the highly controlled conditions of the UHV system.

9. It has been shown that, in terms of sensitivity and speed of response, the presence of oxygen does not introduce reversibility, reproducibility and drift problems when normalized signals are monitored. The sensor has been shown to be safe—in fact, even mixtures including the explosive ratio have been safely checked (there are no ignition sources in PPE detection). The sensitivity of 1000 ppm in O_2 under simulated dynamic flow-through ambient air conditions holds great promise for the establishment of a continuously monitoring H_2 sensor working at room temperature and in open atmospheric air, including the outdoors.

10. Several conclusions have been drawn concerning signal-to-noise ratio optimization and response enhancement of the PPE hydrogen device. Three kinds of signal drift were identified: laser drift, background drift, and hydrogen-induced drift.

11. A Pd–PVDF junction model has been developed based on the photo-thermal response of the pyroelectric. It has been shown that detailed photothermal electrostatic theory is consistent with either, or both, of two operating mechanisms for H_2 gas detection: one involving pyroelectric coefficient dependence on the absorbed hydrogen charge density at the junction, and the other involving the contact potential shift in the presence of hydrogen as a result of thermal-wave modulation of the [H]-dependent Pd work function. The actual operating mechanism appears to depend on the experimental conditions (continuous or intermittent H_2 gas flow). The theory is able to account for the behavior of non-PE photothermal sensors of the type fabricated by Balasubramanian et al. (1991), and it is in good agreement with low T measurements.

REFERENCES

Abe, S., and Hosoya, T. (1984). *Proc. World Hydrogen Energy Conf. 5th*, Toronto, *1984*, Vol. 4, p. 1893.

Aihara, M., and Kawakami, S. (1985). *Proc. Sens. Symp. 5th, IEE Jpn.*, p. 81.

Armgarth, M., and Nylander, C. (1982). *IEEE Electron Devices Lett.* **EDL-3**, 384.

Balasubramanian, A., Santiago-Aviles, J.J., and Zemel, J.N. (1991). *J. Appl. Phys.* **69**, 1102.

Baumann, T., Dacol, F., and Melcher, R.L. (1983). *Appl. Phys. Lett.* **43**, 71.

Bergman, J.G., Crane, G.R., Ballman, A.A., and O'Bryan, H.M., Jr. (1972). *Appl. Phys. Lett.* **21**, 497.

Broadhurst, M.G., and Davis, G.T. (1987). *Top. Appl. Phys.* **33**, 285.

Buccur, R.V., Mecea, V., and Indrea, E. (1976). *J. Less-Common Met.* **49**, 147.

Burns, G., Dacol, F.H., and Melcher, R.L. (1983). *J. Appl. Phys.* **54**, 4228.

Choi, S.-Y., Takahashi, K., Esashi, M., and Matsuo, T. (1986). *Sens. Actuators* **9**, 353.

Christofides, C., and Mandelis, A. (1989a). *J. Appl. Phys.* **66**, 3975.

Christofides, C., and Mandelis, A. (1989b). *Proc. IEEE Ultrason. Symp.*, Montreal, *1989*, Vol. 1, 613.

Christofides, C., and Mandelis, A. (1990). *J. Appl. Phys.* **66**, R1.

Christofides, C., and Mandelis, A. (1991). *Int. J. Hydrogen Energy* **16**, 557.

Christofides, C., Ghandi, K., and Mandelis, A. (1990). *Meas. Sci. Technol.* **1**, 1363.

Christofides, C., Mandelis, A., and Enright, J. (1991). *Jpn. J. Appl. Phys.* **30**, 2916.

Christofides, C., Mandelis, A., Rawski, J., and Rehm, S. (1993). *Rev. Sci. Instrum.* (in press).

Conrad, H.J., Ertl, G., and Latta, E.E. (1974). *Surf. Sci.* **41**, 435.

Cooper, J. (1962). *Rev. Sci. Instrum.* **33**, 92.

Coufal, H.J. (1984). *Appl. Phys. Lett.* **44**, 59.

Coufal, H.J. Grygier, R.K., Horne, D.E., and Fromm, J.E. (1987). *J. Vac. Sci. Technol.* **A5**, 2875.

D'Amico, A., and Zemel, J.N. (1985). *J. Appl. Phys.* **57**, 2460.

D'Amico, A., Palma, A., and Verona, E. (1982). *Proc. IEEE Ultrason. Symp.*, San Diego, *1982*, p. 308.

D'Amico, A., Fortunato, G., Hua, W.G., and Zemel, J.N. (1985a). In *Tech. Dig. 3rd Int. Conf. Solid State Sens. Actuators.*, p. 239. IEEE, Philadelphia.

D'Amico, A., Fortunato, G., Reihua, W., and Zemel, J.N. (1985b). *IEEE Int. Conf. Solid State Sens. Actuators, Transducers '85*, New York, p. 239.

Dannetun, H.M., Petersson, L.-G., Söderberg, D., and Lundström I. (1984). *Appl. Surf. Sci.* **17**, 259.

Faria, I.F., Jr., Ghizoni, C.C., Miranda, L.C.M., and Vargas, H. (1986). *J. Appl. Phys.* **59**, 3294.

Fortunato, G., Bearzotti, A., Caliendo, C., and D'Amico, A. (1989). *Sens. Actuators* **16**, 43.

Frederick, J.R., Zemel, J.N., and Goldfine, N. (1985). *J. Appl. Phys.* **57**, 4936.

Hadni, A. (1963). *J. Phys.* **24**, 694.

Hadni, A. (1980). In *Infrared and Millimeter Waves* (K.J. Button, ed.), Vol. 3, Chapter 3. Academic Press, New York.

Hadni, A., Thomas, R., Magnin, J., and Bagard, M. (1978). *Infrared Phys.* **18**, 663.

Hall, J.P., Whatmore, R.W., and Ainger, F.W. (1984). *Ferroelectrics* **54**, 211.

Hesketh, P., Gebhart, B., and Zemel, J.N. (1985). *J. Appl. Phys.* **57**, 4944.

Hughes, R.C., Schubert, W.K., Zipperian, T.E., Rodriguez, J.L., and Plut, T.A. (1987). *J. Appl. Phys.* **62**, 1074.

Jewett, D.N., and Makrides, A.C. (1965). *Trans. Faraday Soc.* **61**, 932.

KYNAR (1983). *Piezo Film Technical Manual.* Pennwalt Corp., King of Prussia, PA.

Lang, S.B. (1962). *Temp. Meas. Control Sci. Ind.*, **3**, 1015.

Lang, S.B. (1974). *Ferroelectrics* **7**, 231.

Lewis, F.A. (1967). *The Palladium Hydrogen System.* Academic Press, London and New York.

Lundström, I. (1981). *Sens. Actuators* **1**, 403.

Lundström, I., and Söderberg, D. (1981/82). *Sens. Actuators* **2**, 105.

Lundström, I., Shivaraman, M.S., and Svensson, C.M. (1975a). *J. Appl. Phys.* **46**, 3876.

Lundström, I., Shivaraman, M.S., and Svensson, C., and Lundkvist, L. (1975b). *Appl. Phys. Lett.* **26**, 55.

Lundström, I., Shivaraman, M.S., and Svensson, C. (1977). *Surf. Sci.* **64**, 497.

Lundström, I., Armgarth, M., Spetz, A., and Winquist, F. (1986). *Sens. Actuators* **10**, 399.

Lundström, I., Armgarth, M., and Petersson, L.-G. (1989). *CRC Crit. Rev. Solid State Mater. Sci.* **15**, 201.

Lynch, J.F., and Flanagan, T.B. (1973). *J. Phys. Chem.* **77**, 2628.

Mandelis, A. (1984). *Chem. Phys. Lett.* **108**, 388.

Mandelis, A., and Christofides, C. (1989). *Springer Ser. Opt. Sci.* **62**, 347.

Mandelis, A., and Christofides, C. (1990a). *Sens. Actuators* **2**, 79.

Mandelis, A., and Christofides, C. (1990b). *J. Vac. Sci. Technol.* **A8**, 3980.

Mandelis, A., and Christofides, C. (1991). *J. Appl. Phys.* **70**, 4496.

Mandelis, A., and Power, J. (1989). *Appl. Opt.* **27**, 3397.

Mandelis, A., and Zver, M.M. (1985). *J. Appl. Phys.* **57**, 4421.

Mandelis, A., Boroumand, F., Solka, H., Highfield, J., and Van Den Bergh, H. (1990). *Appl. Spectrosc.* **44**, 132.

Marinelli, M., Zammit, U., Mercuri, F., and Pizzoferrato, R. (1992). *J. Appl. Phys.* **72**, 1096.

Mieszkowski, M., Leung, K.F., and Mandelis, A. (1989). *Rev. Sci. Instrum.* **60**, 306.

Mopsik, F.I., and Broadhurst, M. (1975). *J. Appl. Phys.* **46**, 4204.

Munidasa, M., and Mandelis, A. (1991). *J. Opt. Soc. Am.* **A8**, 1851.

Okuyama, M., Togami, Y., Hamakawa, Y., Kimata, M., and Uematsu, S. (1989). *Sens. Actuators* **16**, 263.

Ploss, B., Lehmann, P., Schopf, H., Lessle, T., Bauer, S., and Thieman, U. (1990). *Ferroelectrics* **109**, 223.

Polla, D.L., White, R.M., and Mueller, R.S. (1985). In *Tech. Dig. Int. Conf. Solid State Sens. Actuators*, p. 33. IEEE, Philadelphia.

Putley, E.H. (1970). In *Semiconductors and Semimetals* (R.K. Willardson and A.C. Beer, Eds.), Vol. 5, Chapter 6. Academic Press, New York and London.

Rahnamai, H. (1982). *Sens. Actuators* **3**, 17.

Rahnamai, H., and Zemel, J.N. (1981/82). *Sens. Actuators* **2**, 3.

Rahnamai, H., and Zemel, J.N. (1983). *Sens. Actuators* **3**, 203.

Roentgen, W.C. (1914). *Ann. Phys. (Leipzig)* [4] **45**, 737.

Roundy, C.B., Byer, R.L., Phillion, D.W., and Kuizenga, D.J. (1974). *Opt. Commun.* **10**, 374.

Simpson, J.A., and Tuzzolino, A.J. (1984). *Phys. Rev. Lett.* **52**, 601.

Steele, M.C., Hile, J.W., and MacIver, B.A. (1986). *J. Appl. Phys.* **47**, 2357.

Stillwell, P.F.T. (1981). *J. Phys. E.* **14**, 1113.

Szarka, F.H. (1988). *Fiber Integr. Opt.* **8**, 135.

Vannice, M.A., Benson, J.E., and Boudart, M. (1970). *J. Catal.* **16**, 348.

Wagner, C. (1932). *Z. Phys. Chem., Abt. A* **159**, 459.

Wangsness, R.K. (1986). *Electromagnetic Fields*, Chapter 10. Wiley, New York.

Watton, R. (1976). *Ferroelectrics* **10**, 91.

Whatmore, R.W. (1986). *Rep. Prog. Phys.* **49**, 1335.

Whatmore, R.W., and Bell, A.J. (1981). *Ferroelectrics* **35**, 155.

Yeou, T. (1938). *C.R. Acad. Sci.* **207**, 1042.

Yokoo, T., Shibata, K., and Kuwano, Y. (1986). *Jpn. J. Appl. Phys.* **24**, Suppl. 24-2, 149–152.

Young, J.C. (1982). M.S. Thesis, Dept. Electr. Eng., University of Pennsylvania, Philadelphia.

Zemel, J.N. (1985). In *Solid State Chemical Sensors* (J. Janata and R.J. Huber, Eds.), Chapter 4. Academic Press, New York.

Zemel, J.N. (1990). *Rev. Sci. Instrum.* **61**, 1579.

Zemel, J.N., Keramati, B., Spivak, C.W., and D'Amico, A. (1981). *Sens. Actuators* **1**, 427.

CHAPTER

9

FUTURE TRENDS

Forecasting future developments in the device physics and chemistry of solid state gas sensors is at best an exercise in educated guessing, as trends in the research community can be very much influenced by new breakthroughs and by the emergence of new devices based on conventional or novel principles. It is much easier to predict the market economics of solid state sensors and to forecast the degree of market penetration by members of existing device families in the foreseeable future, based on the existing level of industrial utilization of a given sensor (Tofield, 1987a).

Nevertheless, the knowledge base on device families presented in this volume allows a certain degree of speculation as to the probable directions that the device science of gas sensors may follow in the next 5-10 years.

a. *Semiconductor sensor development* is going to be intimately dependent on further progress in microelectronic material and device processing. The Schottky barrier junction technology is already quite mature, yet more research is required to produce well-behaved sensor devices with reproducible low-tolerance calibration curves. The main problems here are the present-day inability to effectively control interface state densities at the metal–semiconductor junction, which tend to control the position of the Fermi level. Higher silicon substrate purity (at an increased cost), better contacts (i.e., energy-level matching) at the metal–semiconductor interface (Kröger et al., 1956), and standardized wafer cutting and polishing practices will be essential to ensure low variance in the electronic behavior of Schottky barrier sensors. MIS and MiS structures have been studied most intensely to-date (Lundström et al., 1989), yet the main problem that will probably remain unresolved for some time is the drift associated with long-term redistributions in interface charge states. This problem is endemic to capacitive and transistor structures, which involve SiO_x layers and may lead to false alarms following baseline level drift if the device is implemented as a gas sensor. Surface film devices such as SnO_2 sensors (Williams, 1987), based on carrier defect chemistry for conduction at the gas–solid interface, must improve their reliability as well. Much progress has been made in recent years in solid state chemistry (Kofstad, 1972; Tofield, 1977b), which can directly benefit the quality of surface film

309

semiconductor sensors. Possible improvements can be expected to emerge once GaAs microelectronic devices become more mainstream (Einspruch and Wisseman, 1985), with distinct advantages in speed of response.

b. *Photonic and photoacoustic sensors* are still very much at the laboratory stage, with the exception of the portable CO_2-laser-based photoacoustic spectrometer introduced by Sigrist (1992). Further developments appear likely in the area of biological sensing using photoacoustics, such as the plant gas exchange sensor developed by Oehler and Blum (1990). The ultrasensitivity of photoacoustic gas detection is not expected to help this technology displace existing sensors, but rather to establish highly specialized instruments to monitor poisonous and/or lethal gases in trace concentrations in industrial plants where such gases are occluded or used (e.g., semiconductor industry processing laboratories). Photonic sensors based on spectroscopic detection are among the most sophisticated instrumentation-intensive gas monitoring devices and not appropriate for general use. Any further research in this area will most likely be motivated within the academic community, and the prospects in terms of the physics of gas sensing appear to be truly exciting, owing to the unparalleled ability of spectroscopic techniques to probe facets of the interaction mechanisms between gases and surfaces with high specificity as regards the details of the interaction at the molecular or electronic level.

c. *Fiber-optic sensors* form a special group of photonic gas sensors, owing to the existence of mature fiber-optic technologies on which the device implementation of such sensors is based (Measures, 1984; Iizuka, 1985). Although high sensitivity levels have been achieved in detecting specific gases by specially coated optical fibers such as the Mach–Zehnder interferometer used for hydrogen gas detection by Butler (1984), it is not clear that this sensor technology will develop into reliable sensing devices, because of difficulties in calibration and, more importantly, the coexisting optical fiber refractive index sensitivity to factors other than the presence of the monitored gas. These factors primarily include the effects of device sensitivity to small fluctuations in ambient temperature and/or pressure (Hocker, 1979).

d. *Piezoelectric quartz crystal microbalance sensor* applications will be limited owing to the drawback of compromised selectivity, which is related to the resonant adatom–mass sensing mechanism of these sensors. Opposing shifts in the resonant frequency of the PQCMB seem to limit its use as a H_2-gas sensor to the concentration range above 0.4% hydrogen in nitrogen (Christofides and Mandelis, 1989). When coupled to the ambient-temperature-dependent resonance shift of this device, the foregoing considerations present a serious impediment to future developments of the PQCMB into a reliable, quantitative gas sensor technology, despite the simplicity of the relevant instrumentation.

e. *Surface acoustic wave sensors* appear to hold excellent promise of reliability and versatility. Their high sensitivity to trace gas concentrations is due to the nature of SAW, which optimally involves the surface of the device, i.e., the very region where the gas–solid interaction occurs. For that very reason the gas selectivity of SAW structures is excellent (Ricco et al., 1985). Improvements can be expected alongside new progress in thin-film deposition technologies, which will enhance the coupling of the propagating acoustic wave to the active surface layer. Nevertheless, there still persist several uncertainties associated with temperature stability, reproducibility, drift, and interference, as well as device complexity and cost (Tofield, 1987a).

f. *Pyroelectric and thermal sensors* will likely survive into the future in the ac operational mode, which makes them relatively insensitive to drifts common to other gas sensor technologies (Mandelis and Christofides, 1990). Inexpensive and replaceable active elements such as PVDF may become attractive from the point of view of the extended lifetime of the sensor. Minimal baseline drift is an additional advantage, since no interface state charging is likely, unlike the semiconductor-based devices. The well-developed PVDF masking and metallization technology (KYNAR, 1983) raises the possibility of integrated device structures for monitoring several gases simultaneously. This trend is apparent in many gas sensor technologies, but based on the material presented in this volume we believe that only those which depend on well-developed integration methods, notably semiconductor, piezoelectric, and pyroelectric technologies, will be capable of making the transition to multiplexed systems with some assurance of success. It is this promising trend toward integration that will likely sustain the substantial international research efforts in this exciting discipline in the foreseeable future.

REFERENCES

Butler, M.A. (1984). *Appl. Phys. Lett.* **45**, 1007.

Christofides, C., and Mandelis, A. (1989). *J. Appl. Phys.* **66**, 3986.

Einspruch, N.G., and Wisseman, W.R. (1985). *GaAs Microelectronics*, Vol. 11. Academic Press, Ser. VLSI Electron. Microstruct. Sci., Orlando, FL.

Hocker, G.B. (1979). *Appl. Opt.* **18**, 1445.

Iizuka, K. (1985). *Engineering Optics*, 2nd ed. Springer-Verlag, Heidelberg.

Kofstad, P. (1972). *Non-stoichiometry, Diffusion and Electrical Conductivity in Primary Metal Oxides*. Wiley, New York.

Kröger, F.A., Diemer, G., and Klasens, H.A. (1956). *Phys. Rev.* **103**, 279.

KYNAR (1983). *Piezo Film Technical Manual*. Pennwalt Corp., Kind of Prussia, PA.

Lundström, I., Armgarth, M., and Petersson, L.-G. (1989). *CRC Crit. Rev. Solid State Mater. Sci.* **15**, 201.

Mandelis, A., and Christofides, C. (1990). *Sens. Actuators* **2**, 79.

Measures, R.M. (1984). *Laser Remote Sensing.* Wiley, New York.

Oehler, O., and Blum, H. (1990). *Springer Ser. Opt. Sci.* **63**, 369.

Ricco, A.J., Martin, S.J., and Zipperian, T.E. (1985). *Sens. Actuators* **8**, 319.

Sigrist, M.W. (1992). In *Principles and Perspectives of Photothermal and Photoacoustic Phenomena* (A. Mandelis, Ed.), Chapter 7. Elsevier, New York.

Tofield, B.C. (1987a). In *Solid State Gas Sensors* (P.T. Moseley and B.C. Tofield, Eds.), Chapter 10. Adam Hilger, Bristol.

Tofield, B.C. (1977b). In *Reactivity of Solids* (J. Wood, Ed.), p. 253. Plenum, New York.

Williams, D.E. (1987). In *Solid State Gas Sensors* (P.T. Moseley and B.C. Tofield, Eds.), Chapter 5. Adam Hilger, Bristol.

HYDROGEN GAS DETECTORS

In this appendix we review the experimental performance of solid state hydrogen detectors under flow-through conditions available to date such as thermal, pyroelectric, piezoelectric (bulk and surface acoustic waves), photonic and photoacoustic, fiber-optic, and semiconductor hydrogen detectors. A useful feature of this review is a comparison of operating characteristics of each device in Table A.1 (Christofides and Mandelis, 1990).

Table A.1 contains a global comparison of characteristic parameters of several hydrogen detectors presented in this volume. Information is presented on the year of development of each type of detector, the thickness of the catalyst (invariably Pd layer) coated on the device, the operating temperature, the carrier gas under which the experiments were performed, and the sensor sensitivity limit. The response time of the detectors has not been reported in the table because (i) many authors have defined the response time in different ways; (ii) it depends on the volume of the test cell; and (iii) it also depends on the flow rate. The main points of Table A.1 can be summarized as follows:

a. The majority of H_2 sensors (presented in this book) use a Pd metal trap.

b. The oxidation of the palladium layer introduces significant problems for the operation of the hydrogen detectors.

c. The MOSFET hydrogen sensors have the longest history of research and development. This is one of the reasons that so much knowledge concerning their operating mechanism has been accumulated over the last 20 years.

d. MOSFET devices usually operate at high temperature; otherwise their sensitivity and response time are adversely affected. On the other hand, the non-FET devices usually operate under room temperature conditions.

e. At high temperature the FET devices present sensitivities at least 3 orders of magnitude greater than the best non-FET sensor.

f. At room temperature, the highest sensitivity has been exhibited by the Pd–FOS device. We note the enormous progress that has been achieved by the Pd–FOS device. In four years the all-optical sensor became

Table A.1. A Global Comparison of Solid State Sensors under Flow-Through Conditions

References	Type	$L(Å)^a$	$T(°C)$	H_2 in....	Sensitivity
Hall et al. (1984)	Pt–LiTaO$_3$	?	20	Air	300 ppm
Zemel (1985)	Pd–LiTaO$_3$	3000	20	N$_2$	1%
Christofides and Mandelis (1989a) Mandelis and Christofides (1990)	Pd–PPE	130	20	N$_2$	40 ppm
Christofides et al. (1993)	Pd–PPE	1588	20	Air	0.1%
Balasubramanian et al. (1991)	Pd–Mica–Au	?	20	N$_2$	2.5%
Bucur et al. (1976)	Pd–PQCMB	?	81	N$_2$	350 ppm
Abe and Hosoya (1984)	Pd–PQCMB	?	20	N$_2$	0.5%
Christofides and Mandelis (1989b)	Pd–PQCMB	800	20	N$_2$	0.5%
D'Amico et al. (1982/83)	Pd–SAW	3000	20	N$_2$	50 ppm
Butler (1984)	Pd–FOS	?	20	N$_2$	0.2%
Butler (1991)	Pd–FOS	15000	20	N$_2$	20 ppb
Butler and Ginley (1988)	Pd–FOS	?	20	Air	0.1%
Kumar and Fray (1988)	Pd–HUP	—	20	?	100 ppm
Miura et al. (1987)	4–Probes	—	25	Air	0.2%
Lundström et al. (1975a)	Pd–MOS	100	150	Air	40 ppm
Lundström et al. (1975b)	Pd–MOS	10	100	Air	40 ppm
Armgarth et al. (1982)	Pd–MOS	?	150	Air	1 ppm
Armgarth et al. (1982)	Pd–MOS	?	150	N$_2$	0.03 ppb
Armgarth and Nylander (1981)	Pd–MOS	?	150	O$_2$	10–100 ppm
D'Amico et al. (1983)	Pd–MIS	350	22	N$_2$	100 ppm
Steele and MacIver (1976)	Pd–diode	800	25	N$_2$	100 ppm
Furtunato et al. (1989)	Photonic	?	20	N$_2$	1000 ppm
Adler-Golden et al. (1992)	Photonic	?	20	N$_2$	100 ppm

^{a}? = No data available/reported.

more than 8 orders of magnitude more sensitive than the first generation device. Thus, the Pd–FOS is actually the most sensitive detector in the entire group of solid state H_2 sensing devices at room temperature.

g. The photopyroelectric sensor is the second best in terms of sensitivity in the group of non-FET devices even though it has only been developed very recently. The ac PPE device has been found to be almost 3 orders of magnitude more sensitive than the dc pyroelectric sensor.

h. Operation at high temperature of the piezoelectric crystal microbalance leads to a sensitivity 15 times better than that at room temperature. Operation at low H_2 concentrations is, however, impeded by interference from other ambient gases.

i. The SAW device is much more sensitive than the bulk PQCMB sensor.

j. The MOSFET devices have excellent sensitivity even when the carrier gas is not inert. For example, the Pd–MOS device is able to detect up to 10 ppm H_2 even when the carrier gas is oxygen.

REFERENCES

Abe, S., and Hosoya, T. (1984). *Proc. World Hydrogen Energy Conf.*, *5th*, Toronto, *1984*, Vol. 4, p. 1893.

Adler-Golden, S.M., Goldstein, N., Bien, F., Matthew, M.W., Gersh, M.E., Cheng, W.K., and Adams, F.W. (1992). *Appl. Opt.* **31**, 831.

Armgarth, M., and Nylander, C. (1981). *Appl. Phys. Lett.* **39**, 91.

Armgarth, M., Nylander, C., Sundgren, H., and Lundström, I. (1982). *Proc. World Hydrogen Energy Conf.*, *4th*, Pasadena, *1982*, Vol. 4, p. 1717.

Balasubramanian, A., Santiago-Aviles, J.J., and Zemel, J.N. (1991). *J. Appl. Phys.* **69**, 1102.

Bucur, R.V., Mecea, V., and Flanagan, T.B. (1976). *Surf. Sci.* **54**, 477.

Butler, M.A. (1984). *Appl. Phys. Lett.* **45**, 1007.

Butler, M.A. (1991). *J. Electrochem. Soc.* **138**, L46.

Butler, M.A., and Ginley, D.S. (1988). *J. Appl. Phys.* **64**, 3706.

Christofides, C., and Mandelis, A. (1989a). *J. Appl. Phys.* **66**, 3975.

Christofides, C., and Mandelis, A. (1989b). *J. Appl. Phys.* **66**, 3986.

Christofides, C., and Mandelis, A. (1990). *J. Appl. Phys.* **68**, R1.

Christofides, C., Mandelis, A., Rawski, J., and Rehm, S. (1993). *Rev. Sci. Instrum.* (in press).

D'Amico, A., Palma, A., and Verona, E. (1982/83). *Sens. Actuators* **3**, 31.

D'Amico, A., Fortunato, G., Petrocco, G., and Coluzza, C. (1983). *Sens. Actuators* **4**, 349.

Fortunato, G., Bearzotti, A., Caliendo, C., and D'Amico, A. (1989). *Sens. Actuators* **16**, 43.

Hall, J.P., Whatmore, R.W., and Ainger, F.W. (1984). *Ferroelectrics* **54**, 211.

Kumar, R.V., and Fray, D.J. (1988). *Sens. Actuators* **15**, 185.

Lundström, I., Shivaraman, M.S., and Svensson, C.M. (1975a). *J. Appl. Phys.* **46**, 3876.

Lundström, I., Shivaraman, M.S., Svensson, C.M., and Lundkvist, L. (1975b). *Appl. Phys. Lett.* **26**, 55.

Mandelis, A., and Christofides, C. (1990). *Sens. Actuators* **2**, 1990.

Miura, N., Harada, T., and Yamozoe, N. (1987). *Proc. Symp. Chem. Sensors, Electrochem. Soc.* **87-9**, 163.

Steele, M.C., and MacIver, B.A. (1976). *Appl. Phys. Lett.* **28**, 687.

Zemel, J.N. (1985). In *Solid State Chemical Sensors* (J. Janata and R.J. Huber, Eds.), Chapter 4. Academic Press, New York.

INDEX

317